# Pattern and Process in a Forested Ecosystem

F. Herbert Bormann
Gene E. Likens

# Pattern and Process in a Forested Ecosystem

Disturbance, Development and the Steady State Based on the Hubbard Brook Ecosystem Study

Springer-Verlag
New York Heidelberg Berlin

**F. Herbert Bormann**

School of Forestry and Environmental Studies
Yale University
New Haven, Connecticut 06510, USA

**Gene E. Likens**

Section of Ecology and Systematics
Cornell University
Ithaca, New York 14853, USA



**Library of Congress Cataloging in Publication Data**

Bormann, F. Herbert, 1922–
Pattern and process in a forested ecosystem.

Bibliography: p.
Includes index.
1. Forest ecology—New Hampshire—Hubbard Brook Valley. I. Likens, Gene E., 1935– joint author. II. Title.
QH105.N4B67 574.5′264 78–6015
ISBN 0–387–90321–6



Printed in the United States of America

9 8 7 6 5 4 3 2 1

ISBN 0–387–90321–6 Springer–Verlag New York

ISBN 3–540–90321–6 Springer–Verlag Berlin Heidelberg

# Preface

The advent of ecosystem ecology has created great difficulties for ecologists primarily trained as biologists, since inevitably as the field grew, it absorbed components of other disciplines relatively foreign to most ecologists yet vital to the understanding of the structure and function of ecosystems. From the point of view of the biological ecologist struggling to understand the enormous complexity of the biological functions within an ecosystem, the added necessity of integrating biology with geochemistry, hydrology, micrometeorology, geomorphology, pedology, and applied sciences (like silviculture and land use management) often has appeared as an impossible requirement. Ecologists have frequently responded by limiting their perspective to biology with the result that the modeling of species interactions is sometimes considered as modeling ecosystems, or modeling the living fraction of the ecosystems is considered as modeling whole ecosystems. Such of course is not the case, since understanding the structure and function of ecosystems requires sound understanding of inanimate as well as animate processes and often neither can be understood without the other.

About 15 years ago, a view of ecology somewhat different from most then prevailing, coupled with a strong dose of naiveté and a sense of exploration, lead us to believe that consideration of the inanimate side of ecosystem function rather than being just one more annoying complexity might provide exceptional advantages in the study of ecosystems. To examine this possibility, we took two steps which occurred more or less simultaneously. We developed an ecosystem model that heavily involves inanimate as well as animate processes, and we devised a way of calculating some of the major parameters of that model that hitherto could be estimated only with great difficulty.

Our model, which will be described more fully in the subsequent text, conceives of the terrestrial ecosystem as a delimited part of the biogeochemical cycles of the earth. The ecosystem is an open system; it receives energy from the sun and materials and energy from the biogeochemical cycles. It processes these materials and discharges outputs to the larger cycles. Activities within the ecosystem are largely governed by the nature of inputs, and man can have major effects on the structure and function of ecosystems by advertent or inadvertent effects on inputs. Similarly, man can have major effects on the nature of outputs from ecosystems by the kinds of manipulations he consciously or inadvertently promotes within ecosystems. These outputs enter the larger biogeochemical cycles and have the potential of affecting interconnected ecosystems; in other words, one system's outputs become another's inputs. In this light, inputs and outputs can be seen as linkages transmitting the effects of natural or man-made activities between ecosystems. Not only does our concept of the ecosystem emphasize relationships between systems, but it provides a frame whereby the ecosystem can grow or decline in response to natural or man-made developments imposed upon it. Phenomena such as weathering, nitrogen fixation, or biomass accumulation can be evaluated in an ecosystem context, and species strategies can be studied in relation to temporal and spatial biogeochemical variations.

Our ecosystem model probably would have remained an intellectual curiosity had we not conceived of the "small watershed technique" for measuring input-output relationships. This technique considers a small watertight watershed as the delimited ecosystem. As explained in the text, this permits a fairly precise measurement of both input and output of certain nutrients and it permits the construction of partial nutrient budgets. These budgets, coupled with the measurement of parameters within the ecosystem such as productivity, biomass accumulation, canopy leaching, litter fall, and so forth allow the quantification of more complete nutrient budgets. By budget balancing, often we are able to estimate processes that are otherwise very difficult to estimate such as rock weathering, nitrogen fixation, or sulfur impaction.

Not only does the small watershed technique allow quantification of the structure and functions of natural ecosystems, but once having established base-line biogeochemistry, it is possible to develop a program in experimental ecosystem ecology wherein treatments are imposed on entire watershed-ecosystems and responses of the treated system are compared to undisturbed systems. This approach not only yields considerable information on the effects of the treatments but also allows quantification of some processes occurring in the undisturbed system that are otherwise unmeasureable.

In our first volume, *Biogeochemistry of a Forested Ecosystem*, we presented a detailed examination of the biogeochemistry of an aggrading northern hardwood watershed-ecosystem. The primary consideration was on the physical aspects of hydrologic and nutrient flow through the ecosystem and on hydrologic and nutrient budgets.

This volume has as its major concern the presentation of an integrated view of the structure, functions, and development of the northern hardwood ecosystem. It concentrates on the interrelationships among biogeochemical processes, animate and inanimate structure of the ecosystem, species behavior within the ecosystem, and how these relationships change through time following a perturbation. For biogeochemical information it draws heavily on our first volume, but emphasis is placed on the role of biological processes in controlling destabilizing forces to which every ecosystem is continually subjected.

Not only was it our goal to present an integrated view of ecosystem development, but we thought the text should serve, as well, as a teaching tool for natural scientists interested in the structure and function of ecosystems. To that end, we have made a special attempt to detail the reasoning used to reach pivotal conclusions, to separate conclusions based largely on fact from those based largely on speculation and, in general, to write for the reader interested in the ecology of ecosystems rather than for the ecosystem specialist. This is not to deny that concentrated effort on the part of the reader will be required. After all, the ecology of ecosystems is among the most complex of subjects and inevitably some of that complexity must be reflected in the text.

## Acknowledgments

Any scientist working to develop new knowledge owes an incredible debt to those who came before. We wish to acknowledge four persons whose divergent influences converged in a kind of spontaneous generation to give rise to the Hubbard Brook Ecosystem Study. Eugene P. Odum set the scene with his articulate and enthusiastic advocacy for the ecosystem as a basic unit of study in ecology. Murray F. Buell and Henry J. Oosting with their research and teaching emphasis on succession and climax provided the time frame for the developmental aspects of our study, while Arthur D. Hasler's success with experimental manipulation of lake ecosystems lead us to believe that similar success with our systems would be possible. To this group, we add W. Dwight Billings whose encouragement and gentle hectoring led to the writing of this volume.

The Hubbard Brook Ecosystem Study is the product of the minds, hands, and enthusiasm of scores of people. In our first volume, we acknowledged 151 persons whose contributions, great and small, made that volume and the present volume possible. To that list we now add: A. Bormann, H. Buell, T. Butler, J. Cole, P. Coleman, P. Doering, E. Edgerton, J. Ford, C. Goulden, M. Hall, R. Hall, W. Hanson, R. Harkov, T. Hayes, G. Hendrey, D. Hill, W. McDowell, R. Moore, S. Nodvin, J. Sherman, J. Sloane, and J. Teffer. We are also indebted to L. Auchmoody, J.

Baker, R. Campbell, V. Jensen, V. Johnson, E. Kelso, and R. McDonald of the U.S. Forest Service, and especially to J. Miller of the Yale Library for ideas and information on fire and other subjects. For major contributions to the preparation of the manuscript and editorial assistance, we thank Chris Bormann and Rebecca Bormann. Phyllis Toyryla not only typed the manuscript, but provided vital and enthusiastic assistance throughout the preparation of the manuscript.

For keeping us within modest bounds of scientific rigor, we are indebted to colleagues who generously gave their time to read portions of our manuscript. As a result of insightful criticism, perceptive editing, and timely ideas we think the manuscript improved. We also have been made aware that a few of our cherished conclusions are viewed by some as thickets of muddled thinking. We thank all of the following for trying to keep us on a straight and narrow path: S. Bicknell, J. Eaton, R. Holmes, J. Hornbeck, S. Levin (*in vino veritas*), J. Melillo, D. Ryan, T. Siccama, L. Tritton, and T. Wood. Special thanks are due J. Aber, W. Covington, A. Federer, C. Hall, O. Loucks, P. Marks, W. Martin, W. Niering, S. Pilgrim, K. Reed, G. Whitney, and R. Whittaker who reviewed major sections of the book and made dozens of thoughtful suggestions.

We also acknowledge our debt to the National Science Foundation which has supported our work during the last 15 years and to the U.S. Forest Service for making it possible for us to use the Hubbard Brook Experimental Forest as well as to share in the ideas and data accumulated by scientists of the U.S. Forest Service at Hubbard Brook. We add, however, that the ideas expressed here are not necessarily those of the U.S. Forest Service. Finally, space is too short to adequately recognize the overall contributions of Robert S. Pierce. Not only has he been our colleague in the design and execution of science, but major credit for the success of the Hubbard Brook Ecosystem Study is due to his unobtrusive but extraordinary capacity to coordinate and manage a complex enterprise.

# Contents

# Pattern and Process in a Forested Ecosystem

CHAPTER 1

# The Northern Hardwood Forest: A Model for Ecosystem Development

A. S. Watt in his classic paper, "Pattern and process in the plant community" (1947), isolated a central dilemma of modern ecology.

> . . . clearly it is one thing to study the plant community and assess the effects of factors which obviously and directly influence it, and another to study the interrelations of all components of the ecosystem with an equal equipment in all branches of knowledge concerned. With a limited objective, whether it be climate, soil, animals or plants [populations] which are elevated into the central prejudiced position, much of interest and importance to the subordinate studies and . . . to the central study itself is set aside. To have the ultimate even if idealistic objective of fusing the shattered fragments into the original unity is of great scientific and practical importance; practical because so many problems in nature are problems of the ecosystem rather than of soil, animals or plants [populations], and scientific because it is our primary business to understand . . . . [As] T. S. Eliot said of Shakespeare's work: we must know all of it in order to know any of it.

## OBJECTIVES

Even though such a goal remains idealistic today, it is our objective to present an integrated picture of the structure, function, and development over time of the northern hardwood ecosystem in northern New England.

Our ecological studies in the Hubbard Brook Valley within the White Mountains of northern New Hampshire represent a detailed case history of one portion of the northern hardwood forest (Figure 1-1). Over the

**Figure 1-1.** A young second-growth northern hardwood forest at about 600-m elevation on Mount Moosilauke, Warren, New Hampshire. The forest was heavily cut about 65 years ago.

past 15 years these studies have investigated the biogeochemistry and ecology of a variety of forests of different ages. Our basic unit of study has been both natural and experimentally manipulated watershed-ecosystems. These small units of the landscape permit quantitative measurement of biogeochemical input and output. Comprehensive ecological and biogeochemical study of these watershed-ecosystems allows the coupling of animate and inanimate processes to be examined and shows how these internal processes may affect or be affected by the larger biogeochemical cycles of the earth.

We have no grand computerized model where all the animate and inanimate components and processes of the dynamic ecosystem are elegantly linked and where the details of the interactions can be spilled forth by a conversation with the computer. Indeed, no such model exists for any ecosystem. Although we have successfully computerized subcomponents of the northern hardwood ecosystem, our major integration tool is our desire to integrate rather than to dissect. We recognize the enormous limitations to this process in the capability of the human mind, technology, and method. However, even though our analyses are often rudimentary and qualitative or consider only two or three interactive subcomponents, an integrated picture of ecosystem development over time emerges. It is our opinion that this integrated picture provides the most secure base from which we can examine important ecological problems such as the limits to primary production and biomass accumulation, the relationship between species diversity and stability, variations in biogeochemical behavior over time, and the effect of the weathering–erosion interaction on productivity. Most importantly, from this holistic foundation, ecologically sound programs of landscape management can be developed.

Our general philosophy was that, at this stage in the evolution of ecosystem science, a carefully defined and documented case history would provide the best vehicle for generating principles regarding structure and function of an ecosystem. Some principles, rigorously based, could then be tested by ecologists and extrapolated or modified for other ecosystems.

## The Biomass Accumulation Model

Based on measured and projected changes in total biomass, we propose a biomass accumulation model of forest ecosystem development after clear-cutting (Figure 1-2). This model is divided into four phases of development: *Reorganization*, a period of one or two decades during which the ecosystem loses total biomass despite accumulation of living biomass; *Aggradation*, a period of more than a century when the ecosystem accumulates total biomass reaching a peak at the end of the phase; *Transition*, a variable length of time during which total biomass declines; and the *Steady State*, when total biomass fluctuates about a mean.

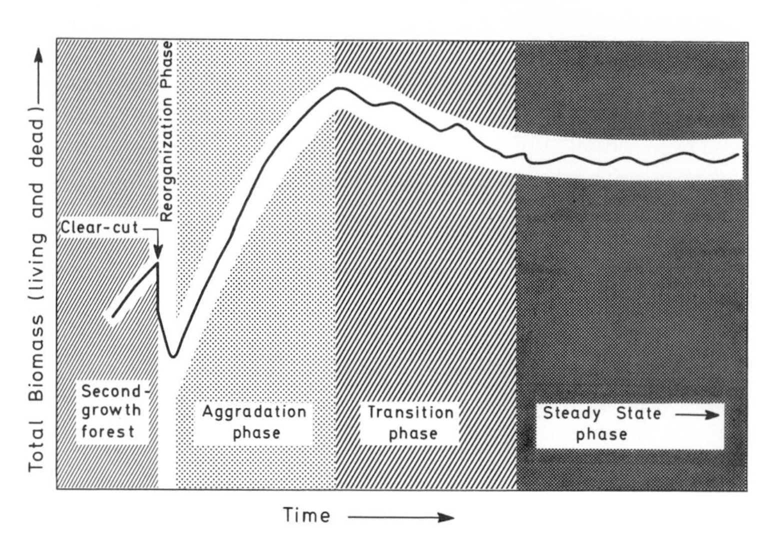


**Figure 1-2.** Proposed phases of ecosystem development after clear-cutting of a second-growth northern hardwood forest. Phases are delimited by changes in total biomass accumulation (living biomass and organic matter in dead wood, the forest floor, and the mineral soil). It is assumed that no exogenous disturbance occurs after clear-cutting.

We define a Steady State in terms of biomass and species composition and biogeochemical function. For the ecosystem *as a whole*, over a reasonable period of time gross primary production equals total ecosystem respiration, and there is no net change in total standing crop of living and dead biomass. Species composition and relative importance of species are fairly constant. However, *any point* within the ecosystem is constantly cycling through changes in both biomass and species composition and function. We use the word Steady State with some trepidation, since ample evidence suggests that the condition we define above is at very best one of slow net changes or possibly a period of quiet antecedent to a cyclic revolution.

Our model of ecosystem development is presented in two parts. Chapters 1 through 5 consider the Reorganization and Aggradation Phases. These chapters contain the greatest degree of realism, since they are based on the wealth of quantitative data accumulated by direct measurement of second-growth forests by the Hubbard Brook Ecosystem Study, the U.S. Forest Service, and other investigators.

Almost all of the present-day forests in northern New England are second-growth forests. They have been cut once or twice within the last century or have been subjected to occasional damage from fire and hurricanes (see Chapter 7). Understanding the behavior of these young forests is thus the key to understanding the theoretical and practical ecology of the present-day northern hardwood ecosystem. In fact, second-growth forests represent the only starting point for an analytical and experimental study, since so-called "climax" forests are extremely rare in New England (Nichols, 1913; Lyon and Bormann, 1961; Bormann and Buell, 1964; Lyon and Reiners, 1971; H. Art, R. Livingston, and T. Siccama, personal communication) or are of questionable history.

The second part of the book, Chapter 6, considers the Transition and Steady State that might develop in the absence of major exogenous (catastrophic) disturbance. This chapter is more hypothetical than earlier chapters since man has virtually eliminated undisturbed old-aged stands in northern New England (see Chapter 7). Chapter 6 draws heavily on the earlier chapters for quantitative aspects of ecosystem dynamics like production, hydrology, and nutrient cycling. Indeed, it is our contention that many aspects of the proposed Steady State are represented in the earlier developmental phases. The Steady-State Phase merely contains spatial diversity that allows for coexistence of conditions that exist earlier along a temporal gradient. Our Steady-State concept, the Shifting-Mosaic Steady State, is based on substantial knowledge of species strategies and behavior developed largely by observational and experimental studies conducted by the U.S. Forest Service and various scientists at Hubbard Brook and by long-term computer simulation of living biomass changes after clear-cutting. Many aspects of the reproductive behavior of the forest associated with later phases of our biomass-accumulation model closely approximate those set forth for forest ecosystems more than a quarter of a century ago in the landmark paper by A. S. Watt (1947).

Ecosystem development may be considered as a battle between the forces of negentropy and entropy or between development of ecosystem organization and its diminishment (Odum, 1969, 1971). Every ecosystem is subject to an array of external energy inputs: radiant energy, wind, water, and gravity. All of these represent potentially destabilizing forces that may destroy or diminish ecosystem organization or sweep away the substance of the ecosystem. For an ecosystem to grow or even maintain itself, it must be able to channel or meet these potentially destabilizing energetic forces in such a way that their full destructive potential is not achieved within the ecosystem.

Some aspects of the control of destabilizing forces in an ecosystem are analogous to a controlled nuclear reaction, in which the enormous energy of the atom is released in small usable amounts rather than in one big bang. Movement of water through an ecosystem, for example, bears many similarities to a controlled chain reaction. Precipitation is first intercepted by the canopy, then by litter on the ground. It is channeled by soil structure through the ground rather than over the ground, and before streamflow from the ecosystem can occur, hydrologic storage capacity must be satisfied. Storage capacity is continually made available by evapotranspiration. Water enters the forested ecosystem with the potential of a lion and, most often, leaves meek as a mouse, with much of its potential energy lost in small frictional increments or simply by conversion of the liquid to vapor by evapotranspiration.

Sometimes incremental dispersion of potential energy is not possible, and structures or strategies (or both) are developed to resist or cope with these destabilizing forces; for example, coastal vegetation is aerodynamically adjusted to strong on-shore winds (Art et al., 1974; Art, 1976), or high-altitude forests maintain themselves in an extremely windy and harsh environment by a strategy of ecosystem development adjusted to wind (Sprugel, 1976; Sprugel and Bormann, 1979).

Control or management of destabilizing forces (e.g., wind, water, and gravity) is the essence of ecosystem development and stability. The success of an ecosystem in resisting destabilization may be judged by its ability to minimize the loss of liquid water and nutrients and to control erosion. Surprisingly, our developmental model indicates that maximum control of destabilizing forces, or maximum "stability," is not the hallmark of the steady-state condition. We shall show that closest control of destabilizing forces is associated with the Aggradation Phase, least control with the Reorganization Phase, and intermediate control with the Transition and Steady-State Phases.

Our model also suggests that ecosystem development is not simply a time when gross primary production exceeds ecosystem respiration. Indeed, there is a fairly substantial period when the reverse is true; for example, the ecosystem has a net loss in total biomass during Reorganization. However, this period is not considered a time of ecosystem degradation but rather a time when the ecosystem "digs deeply" into its nutrient capital to effect rapid repair. Rapid recovery of vegetation tends to minimize erosion and nutrient loss after disturbance. In essence, the ecosystem draws on a bank account of energy and nutrients built up over a long period of time to solve an immediate crisis that threatens still-greater capital losses.

Our proposed Steady State develops in the absence of exogenous disturbance, but relatively small-scale endogenous disturbances operating over long periods of relative quiescence play a major role in its development. Larger-scale exogenous disturbances (e.g., clear-cutting, intense fire, or extensive windthrow) operating over shorter periods of time prevent completion of the development process and drive the ecosystem back to a less developed condition.

## LIMITS FOR OUR THEORETICAL MODEL OF ECOSYSTEM DEVELOPMENT

Our primary goal is to develop, from our detailed case history and other studies, a theoretical understanding, in terms of structure and function, of the temporal relationships of a northern hardwood ecosystem undergoing secondary development. To accomplish this goal, we found it necessary to impose certain limits and to state carefully the assumptions upon which our developmental model is based. As our thinking evolved, four such limits were imposed. These dealt with the geographical extent of the northern hardwood ecosystem, the starting point of development, the role of soil erosion, and the role of exogenous disturbance.

## Relationship of the Ecosystem to the Main Body of the Northern Hardwood Forest

A major question facing us at the outset was, what is the areal extent of the ecosystem to which our findings and hypotheses would apply most directly? To establish some position on this question, we considered the relationship of the northern hardwood forest ecosystem at Hubbard Brook to the main body of forest vegetation of North America. Ideally, it would be more satisfying to establish this relationship by some kind of objective statistical analysis, but a wide-ranging, objective statistical system of vegetational classification awaits development (McIntosh, 1967; Whittaker, 1967). Therefore, we determined the position of our forests within several subjective systems of vegetation classification—systems based on the concept of the community as a discrete, well-defined, integrated unit. Later we will briefly discuss the application of continuum concepts to the northern hardwood ecosystem.

Nichols (1935) includes the Hubbard Brook area in the Hemlock–White Pine Northern Hardwood Region which extends from Nova Scotia to northwestern Minnesota. His classification of forest stands as northern hardwood ecosystems generally rests on the presence of a loosely defined combination of deciduous and coniferous species that may occur as deciduous or mixed deciduous–evergreen stands. Principal species are: *deciduous*–beech (*Fagus grandifolia*) [all scientific names of plant species follow Fernald (1950), except where authorities are cited]; sugar maple (*Acer saccharum*); yellow birch (*Betula alleghaniensis* Britt.); white ash (*Fraxinus americana*); basswood (*Tilia americana*); red maple (*Acer rubrum*); red oak (*Quercus rubra*); white elm (*Ulmus americana*); and *coniferous*–hemlock (*Tsuga canadensis*) red spruce (*Picea rubens*); balsam fir (*Abies balsamea*); and white pine (*Pinus strobus*). Stands classified as northern hardwoods occur under a variety of geologic, topographic, climatic, and historical circumstances.

Within that unit, Braun (1950) would classify northern hardwood forests at Hubbard Brook as red spruce–hardwoods in the New England section of her Northern Appalachian Division. Spruce–hardwood forest is considered to be predominant in northern New England and to grade imperceptibility into boreal spruce–fir forest to the north or at higher elevations and into hemlock–northern hardwoods to the south or at lower elevations.

Oosting (1956) more or less agrees with Braun's interpretation of northern hardwoods but emphasizes the occurrence of boreal forest, mostly at higher elevations, in northern New England (Oosting and Billings, 1951) and shows northern hardwoods extending along the Blue

Ridge Mountains in Virginia and North Carolina (Figure 1-3). Küchler (1964) maps the northern hardwood forest of the United States as extending from Maine to northern Wisconsin and distinguishes four types. The Hubbard Brook area is in Type 108, northern hardwood–spruce forest.

Thus, the northern hardwood forest ecosystem at and around Hubbard Brook is part of an extensive forest type that lies between boreal forest to the north and the main body of the deciduous forest to the south.

The forest at Hubbard Brook also is subject to altitudinal variation. In the mountains of New England and New York one passes through three biomes (deciduous forest, boreal forest, and alpine tundra) with an increase in altitude of a little over 1500-m. This zonation is clearly recognized by students of vegetation and is reflected in a number of regional classification systems (Figure 1-4). However, with the exception of the transition from deciduous northern hardwood forest to evergreen boreal forest at about 760-m, there is little agreement among these systems on units of vegetation or their elevational boundaries. This suggests that the units and their boundaries are rather arbitrarily drawn

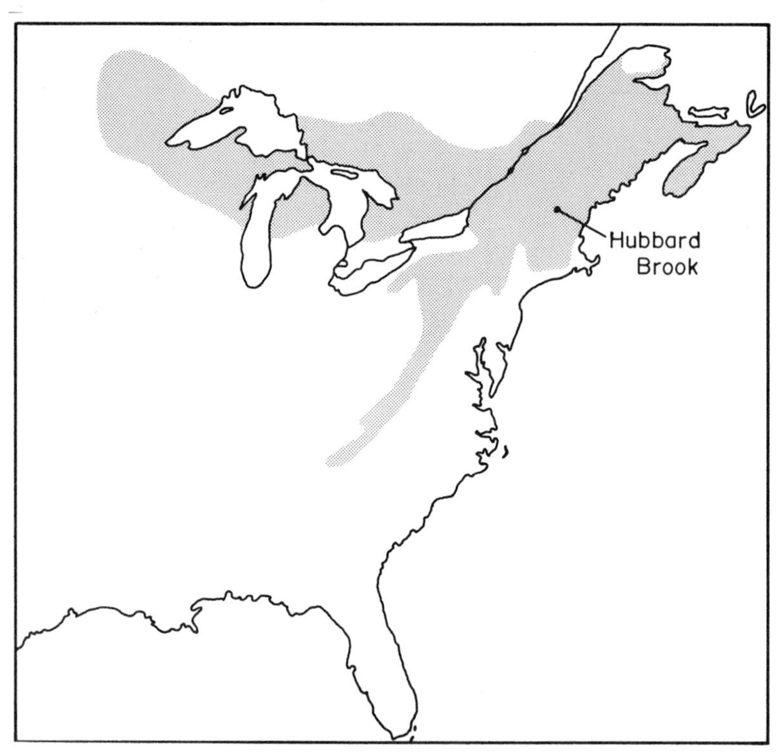


**Figure 1-3.** The approximate range of northern hardwood forest in eastern North America (after Nichols, 1935; Braun, 1950; Oosting, 1956). Areas of boreal forest occur within the eastern part of the area mapped as northern hardwoods but are not delineated.

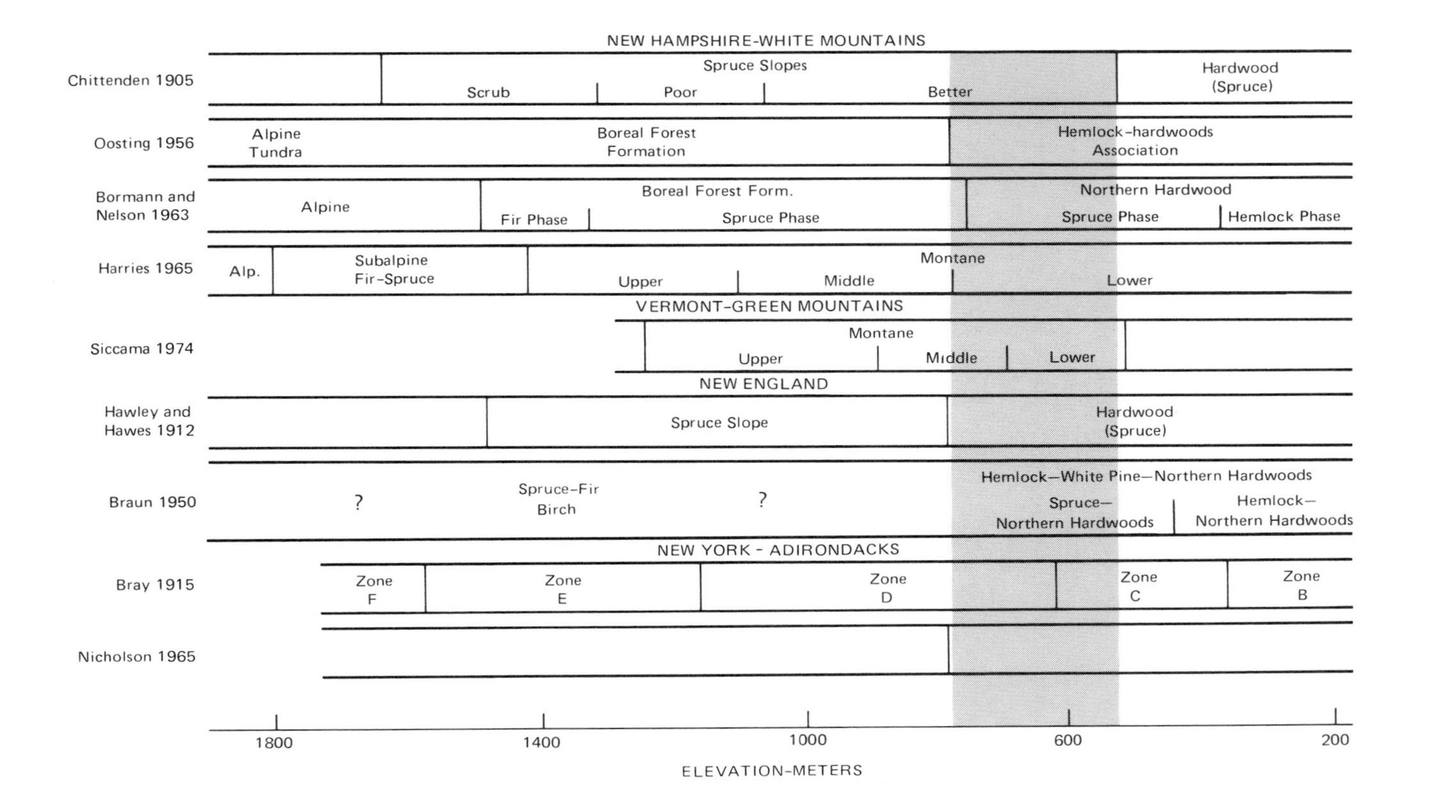


**Figure 1-4.** Classification systems for the vegetation of northern New England and New York State as proposed by various workers. Vegetation zones are plotted against elevation in meters. Shaded area represents that portion of the elevation gradient encompassed by the reference watershed, W6, in the Hubbard Brook Experimental Forest (after Bormann et al., 1970).

and do not reflect discrete community types. In fact, vegetation zones are often defined by the gradual loss or acquisition of species as one moves along an elevational gradient. This situation pictures a complex elevational gradient in which species populations broadly and nonsynchronously overlap along the gradient, which results in a gradual and continuous change in community composition (Whittaker, 1967). Direct gradient analyses in the mountains of New York (Nicholson, 1965) and New Hampshire (Bormann et al., 1970; Leak and Graber, 1974) confirms the independent distribution of tree species along a complex elevational gradient. Observations along a 120-km north–south elevational transect in central New Hampshire also indicate that "successional" as well as "climax" species are distributed in an individualistic pattern (Bormann et al., 1970). Siccama (1974) points out, however, that not all species changes along the elevational gradient are gradual and that the change between northern hardwood forest and spruce–fir forest in the mountains of New England occurs rather abruptly and may reflect a much steeper environmental gradient between 760- and 900-m elevation.

This brief discussion emphasizes that the northern hardwood forest ecosystem at Hubbard Brook is not part of a rigidly defined continental or regional community type but rather is part of a continental vegetation pattern composed of a system of interacting species populations, with the balance between populations shifting in response to changing patterns in the environment and human cultural patterns imposed upon the landscape. This seems to suggest that the concept of the northern hardwood ecosystem is itself untenable. Yet changes in controlling environmental variables or genetic flexibility within species are such that extensive forests of the eastern half of North America are loosely classified by ecologists and foresters as northern hardwood forests. Certainly, field experience throughout northern New England suggests considerable similarity among stands of northern hardwood forests. Until more objective, quantitative means of evaluating relationships between ecosystems arise, we think it useful to retain the concept of a northern hardwood ecosystem bound together by a similarity of vegetation structure, species composition, and ecosystem development processes. However, it is important to recognize that this ecosystem is by no means uniform in biotic and abiotic aspects throughout its geographical range.

To minimize some of the potentially confusing variation in the broad concept of the northern hardwood ecosystem, we defined our system as northern hardwoods occurring within fairly narrow geographical, elevational, and geological limits. Most of our data and ideas are drawn from the ecosystem we have studied most intensively, namely, northern hardwood ecosystems occurring on relatively fine till derived from granite or acid metamorphic rock between elevations of about 250 to 750-m above mean sea level (MSL) in northern New Hampshire. This is roughly equivalent to the area bounded by the White Mountain National Forest. Leak (1978) has classified some of the spatial variation in vegetation occurring within the White Mountains. Almost all of the forests in this area have been cut once or twice, but some of the older forests approximate the maximum living and dead biomass accumulation possible under the prevailing climatic regime.

In later chapters we consider the role of a number of plant species that are generally important within the defined elevational zone. However, we do not consider in detail some species that may be locally important in the lower portion of the zone, such as white pine, hemlock, gray birch, and red oak.

## Initiation of Ecosystem Development

Development of the northern hardwood ecosystem could be initiated in a variety of ways, such as abandonment of agricultural fields, selective or clear-cutting, heavy or light fires, wind storms, or combinations of these. Each initiating cause would introduce variations in the subsequent developmental pattern, particularly in the early stages.

To simplify this situation, we decided to limit our discussion to one starting point, a northern hardwood stand that had been clear-cut. Not only did this provide a uniform starting point, a veritable time zero, but clear-cut areas in all stages of development were available for study within the defined area. We have not only studied commercially clear-cut areas (Pierce et al., 1972; Aber, 1976; Covington, 1976) but also designed, executed, and studied experimental cuttings involving total deforestation, strip-cutting, and controlled commercial cutting (Likens et al., 1970; Pierce et al., 1972; Hobbie and Likens, 1973; Hornbeck, 1973a,b; Bormann et al., 1974; Likens and Bormann, 1974b; Hornbeck, 1975; Hornbeck et al., 1975a,b).

## Soil Erosion

Soil erosion is a major destabilizing force in disturbed ecosystems, and occasionally massive erosion results from sloppy harvesting techniques or poorly planned and constructed roads. Such erosion might drive an ecosystem back to an extremely early developmental stage. Our objective is to examine how an average forest ecosystem develops through time after removal of living forest biomass, *without* massive destruction of biological, hydrological, and nutrient properties of the soil as a direct consequence of the cutting process. Therefore, in fashioning our ideas on secondary development we limited consideration to clear-cut sites with relatively little erosion resulting from mechanical disturbance of the site during the harvesting procedure. Since erosion is of such importance in destabilization, this limitation is basic in that it allows us to evaluate the effect of forest destruction on erosion *in the absence of mechanical disruption*. This has a very practical aspect because it permits the forest manager to evaluate the effects on ecosystem function of cutting alone.

### Exogenous Disturbance

At any time a stand may be subjected to powerful exogenous forces such as wind, harvesting, or fire. These forces are basic elements in determining the structure and function of ecosystems, and they play a major role in shaping the landscape. In many instances, one or all of these forces interrupts the temporal pattern of development; however, we arbitrarily exclude the action of strong exogenous forces and model ecosystem development in the absence of such disturbances. This allows us to examine the role of forces intrinsic to the ecosystem (autogenic forces) in shaping ecosystem development. Later, in Chapter 7, we evaluate the importance of exogenous forces in deflecting our hypothetical pattern of development.

## BIOMASS ACCUMULATION AFTER CLEAR-CUTTING

Flow of biomass within the ecosystem forms a continuum from the moment of formation in gross primary production (GPP) to ultimate dissolution in autotrophic and heterotrophic respiration ($R_A$ and $R_H$) or export from the system. We can think of the biomass of northern hardwood ecosystems as located in five major subcompartments: living biomass in green plants and heterotrophs (animal and plant) and dead biomass in dead wood, the forest floor, and the mineral soil (Figure 1-5).

Woody plant species account for the great bulk of green plants. Heterotrophs, although functionally vital, at any one time constitute a small fraction, <1% (Gosz et al., 1978), of the total living biomass of the ecosystem. The dead wood compartment is composed of branches and trunks on the ground, standing dead trees (Figure 1-6), stumps, and larger dead roots. This material includes bark but is predominantly wood with low nutrient-to-carbon ratios, and is therefore relatively slow to decay. Under some circumstances (e.g., immediately after a clear-cut) the dead wood compartment can represent a sizable portion of the system's biomass. The forest floor is part of the soil profile and is composed of all the organic matter (not including living roots and dead wood) resting on top of the mineral soil (Figure 1-7). Annually the forest floor receives large inputs of relatively easily decomposable fine litter (leaves, twigs, reproductive parts, and fine roots) directly from the living biomass. It also receives input from the dead wood compartment as decay reduces woody material to forms that are incorporated in the soil organic matter. The great bulk of the organic matter in the forest floor is composed of humic compounds which are the products of complex interactions between organic materials and soil organisms. Organic matter in the mineral soil is composed of organic compounds in close physical and chemical relationship with the mineral fraction of the soil. The mineral soil receives organic matter from the forest floor in the form of particulate matter and

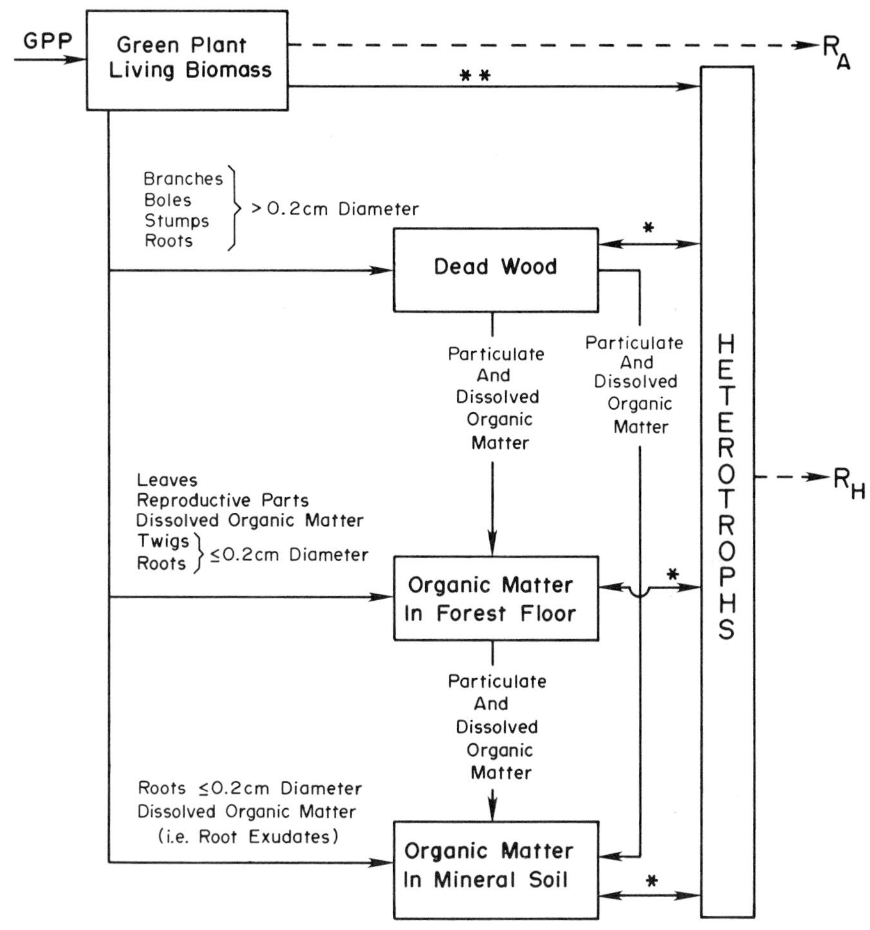


**Figure 1-5.** Biomass distribution within the ecosystem. The location of the standing crop of living and dead biomass *within* the northern hardwood ecosystem and transfers between compartments. Gross Primary Production (GPP) drives the system. Plant and animal heterotrophs represent a very small standing crop but their respiration $R_H$, along with autotrophic respiration $R_A$, is responsible for the ultimate dissolution of biomass within the ecosystem. Arrows to and from the heterotrophic compartment represent the uptake of dissolved and/or particulate organic matter and release resulting from exudation, excretion, fecal material, and death. One asterisk represents activity of saprobes, while two represent activity of green plant parasites and herbivores. For a steady-state ecosystem $GPP = R_A + R_H$, while for a growing system $GPP = R_A + R_H +$ net increase in the sum of the five biomass subcompartments. Input of organic matter in precipitation and output in stream water are not shown in this diagram.

dissolved organic matter and from heterotrophs and roots that decay in place in the mineral soil. The bulk of the humic compounds in the mineral soil is thought to be significantly different chemically from that in the forest floor and presumably has a much longer half-life (Dominski, 1971).

**Figure 1-6.** A large standing dead American elm tree. Note the gap in the overstory canopy created by the death of the crown. Elevation about 550-m. Northern hardwood forest at Gifford Woods State Park, Vermont.

**Figure 1-7.** A spodosol soil (Becket Pedon) at Hubbard Brook. Subdivisions of the B- and C-horizons are mapped but not shown. The forest floor is mostly organic matter with a low bulk density and is usually sharply demarcated from the underlying mineral soil.

In the ecosystem that we follow through time, development is initiated by clear-cutting and the removal of large quantities of biomass. Almost all subsequent biomass originates within the ecosystem from gross primary production (GPP); only a very small amount of organic carbon enters the ecosystem in precipitation and dry fallout (Gosz et al., 1978). Ecosystem biomass is reduced by autotrophic respiration of green plants ($R_A$), by the respiration of heterotrophs ($R_H$), and by export of dissolved organic matter and particulate organic matter in water draining the ecosystem. Generally, export of organic matter in water has been shown to constitute a small part of the total loss (Likens et al., 1977). Throughout, we assume that the importation and exportation of organic matter by animals crossing the boundary of the ecosystem is about balanced (Bormann and Likens, 1967), but during early regrowth after clear-cutting net removal by grazing animals may be a factor of some importance. This matter is currently under study at Hubbard Brook.

Organic matter is physically transferred along a variety of pathways within the ecosystem (Figure 1-5). Transfers of organic matter are made from the green plant compartment to dead wood, forest floor, and mineral soil, as well as to herbivores and parasites. Net primary productivity (NPP) equals the sum of the accumulation of green plant biomass plus any net transfer of organic carbon from the green plant compartment to other compartments within the ecosystem. The whole is summed by the equation $\mathrm{GPP} - R_{\mathrm{A}} = \mathrm{NPP}$.

## Living Plant Biomass

At the inception of the Hubbard Brook Ecosystem Study we assumed that the second-growth forests cut during the period 1909–1917 were more or less phytosociologically stable or close to a "climax" condition in the traditional sense. Vegetational analysis (Bormann et al., 1970; Siccama et al., 1970; Forcier, 1973) confirmed species arrays and population structures that conform well to traditional phytosociological concepts of climax (Nichols, 1935; Braun, 1950; Oosting, 1956). New data (Botkin et al., 1972a,b; Gosz et al., 1973, 1976; Whittaker et al., 1974) indicate that rates of net accumulation of living biomass are relatively high and that on a production basis it is not feasible to consider these forests as approaching steady state, rather they are young, rapidly changing forests. Later in our discussion of steady state, we will show that present-day Hubbard Brook forests do not conform to phytosociological "climax" criteria either.

Whittaker et al. (1974), using dimension analyses, estimated production characteristics of the second-growth forest at Hubbard Brook (Table 1-1). Samples taken in 1966 indicated an abrupt and curious decline in productivity from the period 1956–60 to 1961–65. These workers proposed that the growth period 1956–60 represents the more normal situation, while data from 1961–65 reflect a period of subnormal growth that currently dominates northern New England. We have combined these two periods to approximate an "average" growth period.

**Table 1.1.** Production Characteristics of a 55-Yr-old Second-Growth Forest at Hubbard Brook[a]. Dry Grams per Square Meter per Year.

| | Gross Primary Production (GPP) | | Net Primary Production (NPP) | | Net Living Biomass Accumulation | |
|---|---|---|---|---|---|---|
| Period of Measurement | Above-ground | Total | Above-ground | Total | Above-ground | Total |
| 1956–60 | 2127 | 2547 | 957 | 1147 | 350 | 435 |
| 1961–65 | 1760 | 2090 | 792 | 941 | 238 | 290 |
| Average | 1944 | 2319 | 875 | 1044 | 294 | 363 |

[a] *The forest was selectively cut about 1909 and then heavily cut about 1917. As a matter of convenience we use 55 years, but various ecological measurements have been made over the last 15 years. From Whittaker et al. (1974).*

Net primary production (NPP) for the entire ecosystem is estimated to be 875 g of dry matter/$m^2$-yr aboveground NPP and 1044 g/$m^2$-yr total NPP for the average period. The data (Table 1-1) indicate that for the average period about 363 g/$m^2$-yr, or 35%, of the current NPP is being incorporated as a net increase in living biomass of the ecosystem. The Hubbard Brook forest is thus comparable to other young second–growth temperate deciduous forests (Whittaker et al., 1974).

It is impossible to consider living biomass accumulation over several centuries by a study of a sequence of existing stands: other methods must be used. Whittaker et al. (1974), using 1956–60 estimates of current production and woody litterfall, project a living biomass for the climax condition at Hubbard Brook of about 420 t/ha, 90% of which would be achieved within 250 years after cutting.

We can also estimate long-term living biomass trends using JABOWA, a computer model for northern hardwood forests (Botkin et al., 1972a,b). JABOWA, with some of its data base derived from the Hubbard Brook Ecosystem Study, simulates the growth of individual trees in a mixed species forest on $10 \times 10$-m plots, taking into account competition among trees, response of individual species to environmental variables such as soil moisture, light, and nutrients, introduction of new individuals, and growth and survival characteristics of species. It does not take into account predators or parasites or interplot effects. The basic model keeps track of the number and diameter of stems by species. This has been expanded through the application of dimension analysis (Whittaker et al., 1974) and tissue nutrient concentrations (Likens and Bormann, 1970) to yield changes in total plot living biomass and nutrient content through time.

Using the computer output from 100 plots, we can estimate, on an annual basis, a variety of species and ecosystem parameters (e.g., density, frequency, basal area, biomass, nutrient standing crop) on a hectare basis.

Throughout the text we use JABOWA to predict long-term biomass and nutrient content for a northern hardwood ecosystem on deep till, at an elevation of 600-m, and under average growing conditions. Under these environmental conditions, projections from JABOWA show a peak living biomass of about 410 t/ha about 170 years after clear-cutting.

Although we use JABOWA projections (Figures 1-8 and 1-10), we recognize that any simulation is only as good as the assumptions underlying it and that, for lack of knowledge, some assumptions are rough estimates. To evaluate these projections we used stand data from a number of present-day older northern hardwood forests (Chittenden, 1905; Siccama, 1974; H. Art, personal communication; W. Leak, personal communication) and Whittaker's dimension-analysis regressions (Whittaker et al., 1974) to calculate living biomass. For these stands, living biomass ranged from 325 to 390 t/ha. When compared with these data, both the Whittaker and JABOWA living biomass estimates appear reasonable.

JABOWA has the added advantage that it yields detailed information on annual production and nutrient characteristics by species and by plant parts and for the ecosystem as a whole. These data allow us to make a variety of mass-balance calculations which are utilized throughout the text. Finally, with JABOWA we can produce a 500-yr biomass curve that is quite different from the asymptotic curve projected by Whittaker et al. (1974). We believe that the biomass curve from JABOWA represents a rational index of ecosystem development and we later use it as the basis for our discussion of the steady-state ecosystem.

## Dead Wood Compartment

Until recently, rotting wood in the forest was the special preserve of the entomologist and mycologist hunting specimens; now it is realized that dead wood plays an important role in ecosystem dynamics.

Despite its recognized importance, dead wood is rarely included in biomass estimates, which usually focus on living biomass and organic matter in the soil profile. This void in data on dead wood exists in part because of the great difficulties involved in its estimation. Dead wood occurs in locations difficult to sample such as dead limbs on living trees or standing dead trees. It is often difficult to determine when dead wood ceases to be dead wood and becomes part of another ecosystem compartment; for example, the same fallen log will often contain sound wood and all degrees of decayed wood, including some material properly considered part of the forest floor. Finally, dead wood is probably the least predictable component of the ecosystem because it is greatly influenced by stochastic events like local wind and ice storms which can, at infrequent intervals, add large amounts of previously living biomass to the dead wood compartment.

Dead wood flux is of considerable importance in northern hardwood ecosystems, and we recently instituted an intensive study of dead wood at Hubbard Brook. That study will be completed in several years. In the meantime, we use preliminary data to estimate changes in dead wood during ecosystem development following clear-cutting. We approximated dead wood relationships in several ways.

Observations in a sequence of different-aged successional stands after clear-cutting indicated (1) that a large amount of dead wood (slash) is left immediately after cutting and (2) that dead wood declines to a minimum about two decades after clear-cutting when most of the dead wood present at time zero (clear-cutting) has been respired or transferred to other compartments and before there has been appreciable transfer of new dead wood from the living plant biomass compartment (Figure 1-5). Thereafter, the dead wood compartment once again increases in size.

The amount of woody debris transferred from formerly living trees to dead wood as a result of clear-cutting depends on the age and condition of the stand at the time of cutting and the wood products removed from the ecosystem. We estimated this transfer using data from a 55-yr-old stand with 169 t/ha of above- and belowground living biomass. We computed that 81 t/ha were removed in forest products (saw logs, pulpwood, and millwood) and that 3 t/ha of leaves and twigs were transferred directly to the forest floor. Thus 85 t/ha of slash, stumps, and roots were added to an estimated 36 t/ha of dead wood in the stand at the time of cutting to yield 121 t/ha of dead wood in the ecosystem immediately after clear-cutting.

To gain a rough estimate of the standing crop of dead wood at different times we sampled two stands of different ages: a 56-yr-old stand at the Hubbard Brook Experimental Forest and the Charcoal Hearth Forest, a so-called virgin forest about 20 km from Hubbard Brook (Table 1-2). Using plot techniques, dead wood on the forest floor and in dead stumps with attached roots was collected and its oven-dry weight was obtained. Diameters and heights of standing trees were also measured; biomass content was estimated from the diameter using dimension-analysis regression equations (Whittaker et al., 1974) and then reduced by a fraction derived from the actual height of the dead tree divided by the estimated height for a living tree of that species for that diameter. Since biomass thus estimated is solid wood and bark, we reduced the estimate by 50% to account for decay that had already occurred. In a separate study, J. Roskoski (personal communication) sampled dead wood in 10 × 50-m areas laid out on the forest floor in various-aged stands (Table 1-2). Although both sets of estimates are rough, they show substantial agreement and provide a reasonable guide to the amounts of dead wood that might be expected in various-aged stands.

Dead wood may be lost from the dead wood compartment by oxidation or consumption by heterotrophic organisms and by the transfer of

**Table 1-2.** Estimates of Dead Wood in Northern Hardwood Stands of Various Ages

| | Dead Wood (t/ha) | | |
|---|---|---|---|
| Stand Age (Years) | On or In Forest Floor[a] | Standing | Total |
| 4[b] | 37.7 ± 8.4 | | 37.7 |
| 8[b] | 58.6 ± 31.8 | | 58.6 |
| 18[b] | 6.9 ± 2.1 | | 6.9 |
| 40[b] | 9.1 ± 3.2 | | 9.1 |
| 56[c] | 29.4 ± 5.4 | 4.4[d] | 33.8 |
| 57[b] | 20.9 ± 5.1 | | 20.9 |
| >170[b] | 34.4 ± 17.6 | | 34.4 |
| >170[c] | 28.6 ± 5.3 | 4.9[d] | 28.6 |

[a] *Mean ± standard error of the mean.*
[b] *From Roskoski (1977); based on five 10-m² plots.*
[c] *Based on fifty 1-m² plots.*
[d] *Includes estimate of dead roots.*

partially decomposed organic matter to the forest floor or to the mineral soil (Figure 1-5). We considered these two ways of removal as a single output called the disappearance rate. The disappearance rate of dead wood is difficult to estimate because dead wood includes a variety of plant materials of different sizes and nutrient contents located in various positions aboveground (Figure 1-6) and belowground. To arrive at a rough approximation, we simulated disappearance from the dead wood compartment using assumptions based on Spaulding and Hansbrough's study (1944) of the disappearance of northern hardwood logging slash and on personal observations.

The simulation began with 121 t/ha in the dead wood compartment (slash plus dead wood in the ecosystem at the time of cutting); thereafter, all input of dead wood resulted from tree death as stand development proceeded. Annual additions of dead trees were computed using JABOWA. Input of dead wood resulting from clear-cutting and from the death of trees during development was added to the following classes of materials within the dead wood compartment: (1) roots, (2) tops of trees <20-cm dbh [diameter at breast height], (3) tops of trees >20-cm dbh, (4) trunks of trees <20-cm dbh, and (5) trunks of trees >20-cm dbh. Classes 2 and 3 contained a variety of branches, limbs, and trunks. For all classes we assigned a delayed-exponential disappearance rate such that disappearance starts slowly, speeds up as the rate of decay and transfer to other compartments increases, and then slows as more recalcitrant woody tissues become an increasingly important part of the decay substrate. For roots, Class 1, we assumed that relatively little weight would be lost for 2 years after death and that all but 5% would disappear in 10 years; for Class 2, 3 years of slow loss with 5% remaining in 12 years; Classes 3 and 4, 4 and 15 years; and large trunks, Class 5, 5 and 25 years. Input, output (disappearance), and standing crop were computed annually (Figure 1-8).

The simulator was tuned somewhat to reflect field observations. During the first 20 years after clear-cutting there is a sharp decline in the standing crop of dead wood. Dead wood reaches a low point after about 20 years and then begins to increase. From Year 50 through Year 170, the standing crop fluctuates between 29 and 44 t/ha. In our biomass summary (Figure 1-10), dead wood is represented by a hand-fitted curve based on the rough simulation shown in Figure 1-8.

## The Forest Floor

Dominski (1971) has shown for the experimentally deforested watershed at Hubbard Brook where all regrowth was suppressed that after 3 years the depth of the forest floor had decreased by an average of 3 cm with a loss of 24% of the original weight. Hart (1961) also found that 20 to 30 years after clear-cutting of forest stands in the Hubbard Brook area forest floor depths were as much as 2.5 cm less than estimated original depths. Several major questions arise from these observations. For how long and to what degree does the forest floor undergo a net loss of organic matter

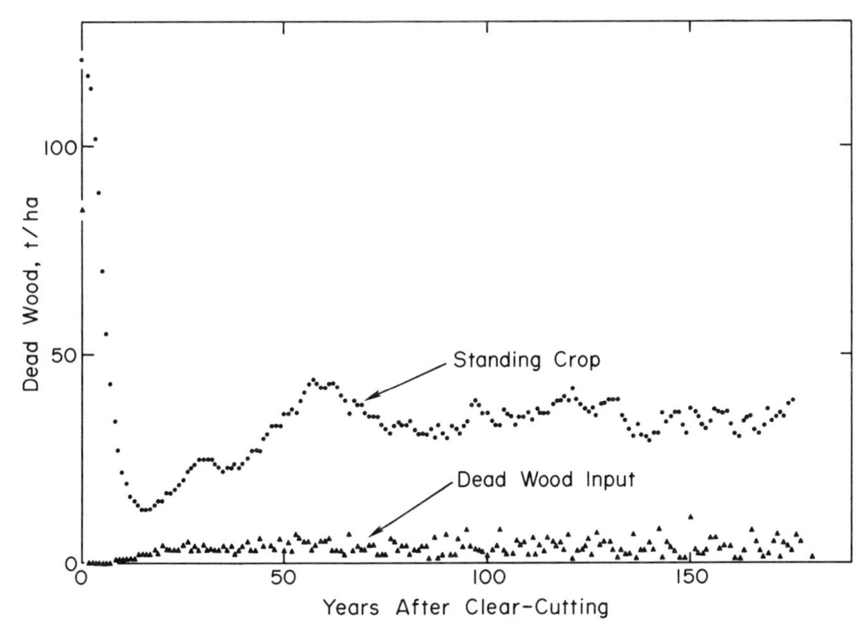


**Figure 1-8.** Simulation of the standing crop of dead wood in northern hardwood forests at various times after clear-cutting. The large quantity of dead wood at the beginning is largely due to slash left from the clear-cutting. Dead wood input after clear-cutting results from the death of trees as predicted by JABOWA, the forest growth simulator.

after clear-cutting, and how and when is this material replaced?

Studies by Morey (1942), Sartz and Huttinger (1950), and Trimble and Lull (1956) indicate that the forest floor under northeastern northern hardwood forests continues to decrease in depth for one to two decades after cutting. Thereafter, it begins to increase in depth, leveling off about 100 years after cutting. These observations point up a remarkably interesting aspect of ecosystem dynamics. Apparently the forest floor continues to lose organic matter even after the cutover area is fully vegetated, litterfall is reinstituted, and conditions of soil temperature and moisture are not greatly different from the precutting conditions. Only after one or two decades does the forest floor begin to regain biomass.

Because these publications on northeastern forests gave little precise information on stand history, local environment, or biogeochemistry, Covington (1976) repeated the work of previous workers (Morey, 1942; Sartz and Huttinger, 1950; Trimble and Lull, 1956). He selected a sequence of 14 northern hardwood stands of known history in the White Mountain region. These stands had been clear-cut at various times, and all met similar site, aspect, geology, and elevation requirements.

Each stand was sampled with thirty 10 × 10-cm plots randomly distributed within a 50 × 100-m rectangle. The forest floor was harvested, oven-dried, and passed through a 2-mm sieve. All living roots, stones, and dead wood that could not be forced through the sieve with modest pressure were removed and the organic matter content of the remainder was obtained by loss-on-ignition.

Covington's data provide a detailed quantification of forest floor behavior after clear-cutting. Organic matter content decreased for about 15 years before reaching a minimum (Figures 1-9 and 1-10). A dry-weight loss of about 31 t/ha, a decline of 51%, occurred during this period. Accretion began about Year 15 and continued asymptotically achieving about 95% of the projected asymptote, 56 t/ha, by Year 65. Determination of the time at which rate of accumulation of forest floor biomass levels off is difficult because of an almost complete lack of older stands with a known cutting history. However, Covington's conclusion of a forest floor asymptote fits with conclusions of other workers (Sartz and Huttinger, 1950; Trimble and Lull, 1956; Lull, 1959; Leak, 1974) based on the study of presumably old-aged stands.

Control of the dynamics of the forest floor, i.e., a net drop in biomass over a decade and a half, while the living biomass is accumulating rapidly, is not yet fully understood. Covington (1976) has suggested that several factors are involved. Immediately after cutting, increased soil temperature, soil moisture, and nutrient availability may speed decomposition. Within only a few years, however, development of dense vegetation produces microclimatic conditions in the forest floor not greatly different than those of the 60-yr-old forest. Covington (1976) suggested that the continued loss of forest floor material may be due to changes in the quality and quantity of litter. That is, living biomass, during the first stage

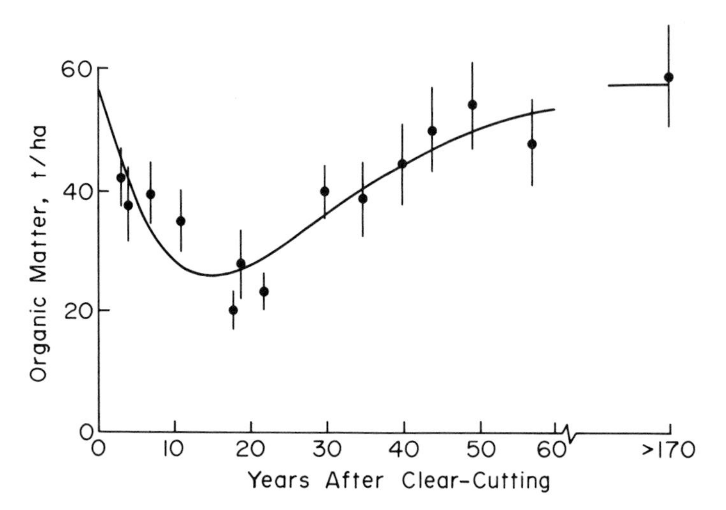


**Figure 1-9.** The weight of organic matter in the forest floor in relation to the age of northern hardwood stands after clear-cutting. The oldest stand, over 170 years, was probably never cut. Points are means of 30 samples per stand with 95% confidence intervals. The curve is a least-squares fit of a gamma function. Organic matter levels are adjusted to eliminate sampling bias (after Covington, 1976).

of ecosystem development after cutting, not only produces less litter but this litter has a higher proportion of leaf to woody material and is generally more decomposable than that of later stages. Leaves of pin cherry, *Prunus pensylvanica*, and raspberry, *Rubus idaeus*, important early developmental species, have higher nutrient content, are softer, and have less fiber content than leaves of species that dominate the ecosystem a little later in development. Covington (1976) proposed that the forest floor shifts from a mode of net loss to one of net gain as the litter becomes more resistant to decomposition and the proportion of wood to leaf litter increases during ecosystem development.

## The Mineral Soil

The largest quantity of organic matter in our second-growth forests is incorporated in the mineral soil (Figure 1-7). We estimate the amount to be about 173,000 kg/ha to a depth of 45-cm in a 55-yr-old forest (calculated from Dominski, 1971). The error associated with this estimate is unknown, but it might be as much as ±20%. The source of error comes from the spatial variability of the soil profile due to rock outcrops, very rocky horizons, and local pit and mound topography.

A net increase or decline of organic matter in the mineral soil after clear-cutting could have a major effect on the biomass balance of the ecosystem as a whole. Such change is difficult to measure, yet because of the large quantity of organic matter in the mineral soil it is important to attempt some evaluation of potential changes in response to clear-cutting. To do so requires a brief discussion of the dynamics of organic matter in the mineral soil.

Organic matter may be added to the mineral soil by several routes (Figure 1-5). Some material may be added by the mixing of organic particulate matter from the forest floor into the uppermost part of the mineral soil, but this route is not well developed since large populations of mixing animals are absent from these soil profiles (Lutz and Chandler, 1946). Generally, there is a fairly discrete disjunction between the lower part of the forest floor and the upper part of the mineral soil (Dominski, 1971). Often the forest floor lies directly on top of a leached $A_2$-horizon (Figure 1-7). Another route of transfer is by addition of decomposition products from dead roots in the mineral soil. Larger roots in the mineral soil constitute a relatively small part of the dead wood compartment since most dead wood is located aboveground and in the forest floor. Fine roots also tend to be concentrated in the forest floor. A major route of organic matter transfer is by translocation of dissolved organic compounds from upper horizons and their precipitation or coprecipitation in the B-horizon of the soil profile (Hurst and Burges, 1967; Stevenson, 1967; Brady, 1974). Recent evidence suggests that some fine particulate matter may be transported to the B-horizon in percolating water (Ugolini et al., 1977).

In terms of flux rates (Figure 1-5), the forest floor is the greatest recipient of dead organic matter from the living biomass compartment. Annual aboveground litter input to the forest floor is 5.7 t/ha. Belowground litter is estimated to be about 1 t/ha, most of which is also deposited in the forest floor (Gosz et al., 1976).

It seems likely that organic matter input into the forest floor is one or two orders of magnitude greater than organic matter input into the mineral soil based on standing crop/input. Dominski (1971) and Gosz et al. (1976) estimate that the standing crop of organic matter in the forest floor at Hubbard Brook turns over, on the average, every 7 or 8 years. Although we have no direct data on turnover times of organic matter in the B-horizon, Hurst and Burges (1967) cite ages of humic materials based on $^{14}C$ analysis of $360 \pm 60$ yr for gray-wooded podzolic soils, $370 \pm 100$ yr for the B-horizon of a Swedish podzol, and 1580 to 2860 yr for the B-horizon under a heathland podzol in East Anglia. There are difficulties with interpretation of $^{14}C$ analysis in soil (O'Brien and Stout, 1978). However, if we assume an average residence time of about 300 years for organic matter in the mineral horizons at Hubbard Brook, this would suggest an average annual input of about 0.6 t/ha (173 t/ha/300 yr = 0.6 t/ha/yr). This input to mineral horizons is about one-tenth of the input estimated for the forest floor.

Although the standing crop of organic matter in the mineral soil is about three times greater than that in the forest floor, it may be considered as far less responsive to environmental change. During the first two decades after clear-cutting one would not expect the organic matter content of the mineral soil to increase reciprocally as the dead wood and forest floor compartments decreased. Some increase might be expected as a result of increased translocation of dissolved substances to the mineral soil or possibly increased mixing of particulate organic matter in upper mineral soil. On the other hand, some decrease in standing crop might occur as a result of increased decomposition due to warmer soils (Dominski, 1971) and an abundance of dissolved nutrients in the mineral soil.

We found it difficult to make a direct measure of short-term changes in the standing crop of dead biomass in the mineral soil because such changes are small in relation to the size of the standing crop and because spatial variability of the soil profile and rockiness of the horizon presents a difficult sampling situation with a relatively high associated error. However, two lines of evidence suggest that the standing crop of dead biomass in the mineral soil does not increase significantly after clear-cutting. For our experimentally deforested watershed where revegetation was prevented for 3 years, approximately 360-kg of nitrogen/ha were flushed out of the ecosystem in stream water (Likens and Bormann, 1974a). At the same time the depth of the forest floor on the watershed decreased about 3 cm with an estimated loss of 470-kg of nitrogen/ha (Dominski, 1971). Although not conclusive, these data indicate that the bulk of the missing forest floor organic matter was mineralized, the end products were flushed out of the ecosystem, and a relatively small proportion of organic matter may have been added to the mineral soil. During this same period, Dominski (1971) measured a decrease in both the nitrogen and organic matter content of the mineral soil, but the loss was relatively small and the estimate was subject to a relatively large error.

Based on the above data, we use the following working hypothesis for dead biomass in the mineral soil after clear-cutting. Biomass in the mineral soil compartment (Figure 1-5) is the end product of relatively slow accumulation over time. It is, in general, less responsive to change, particularly in the B-horizon, than biomass in either the dead wood or forest floor compartments. Immediately after clear-cutting, the standing crop of dead biomass in the mineral soil may increase or decrease in weight, but the net change will be relatively small in relation to the other biomass compartments and will have relatively little effect on the total biomass of the ecosystem.

Given the great difficulty in actually measuring changes in organic matter content of the mineral soil through time, and in view of the arguments just presented, we assume that changes in organic matter content in the mineral soil during development after clear-cutting have only a minor influence on the overall biomass balance of the ecosystem (Figure 1-10). This is probably a reasonably sound assumption for a period of several decades after clear-cutting but is less so over several centuries of time.

## Early Phases of Ecosystem Development Based on Changes in Total Biomass

Given the assumptions in the previous sections, changes in the total standing crop of biomass for the ecosystem over a 170-yr period after clear-cutting can be estimated by summing biomass in the individual compartments (Figure 1-10). A positive slope in the total biomass curve indicates a net accumulation of biomass for the ecosystem as a whole; e.g., GPP is greater than $R_{A+H}$—energy fixed by photosynthesis exceeds that dispersed by total-system (autotrophic + heterotrophic) respiration. A negative slope indicates a net loss of biomass where $R_{A+H}$ is greater than GPP or a situation where total-system decomposition exceeds photosynthesis.

This analysis reveals a number of interesting features about ecosystem functions. The first and most significant aspect of the total biomass curve is that ecosystem behavior after clear-cutting is not characterized by a simple curve for biomass accumulation. For about 15 years after clear-cutting there is a net loss in biomass, and only after that does accumulation occur. The cause of this pattern resides in the nonsynchronous behavior of the standing crops of dead and living organic matter. Both the forest floor and the dead wood compartments show marked initial declines in standing crop, which together outweigh the net accumulation of living

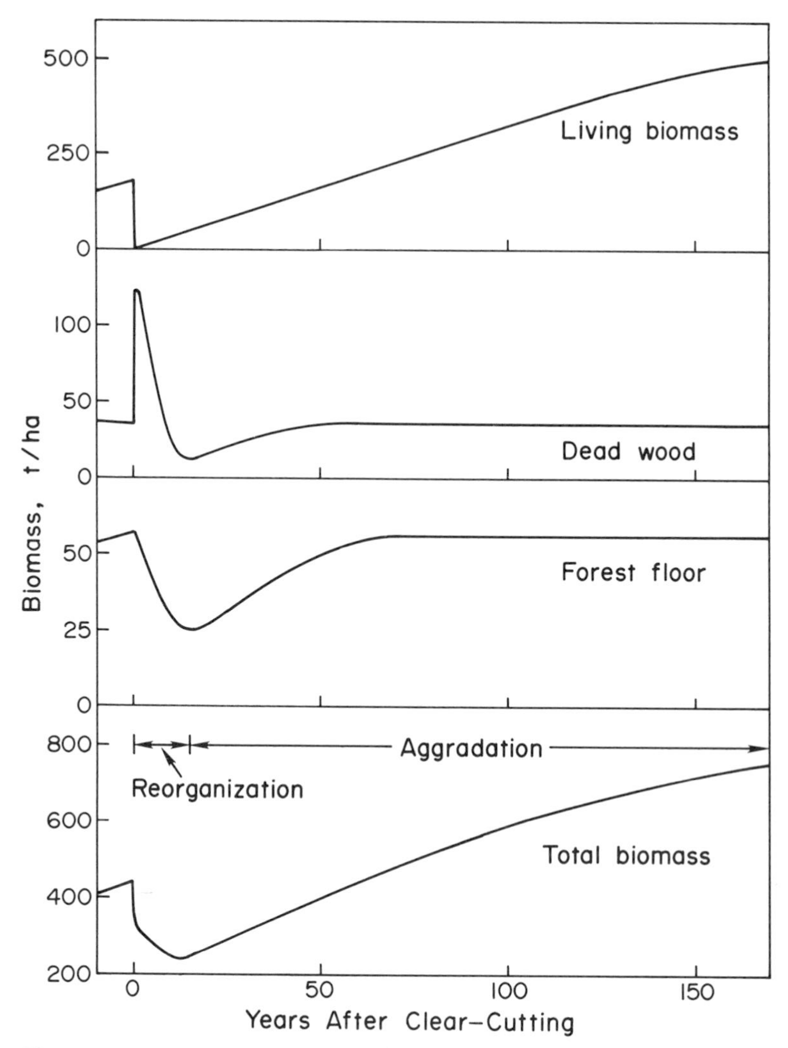


**Figure 1-10.** Biomass summary for a northern hardwood forest following clear-cutting. Living biomass is from a JABOWA simulation, dead wood is from Figure 1-8, and forest floor is from Figure 1-9. Organic matter in the mineral soil, 173 t/ha (after Dominski, 1971), is assumed to undergo little quantitative change. Total biomass is the sum of living biomass and dead organic matter in dead wood, the forest floor, and in the mineral soil. Only the Reorganization and Aggradation Phases of ecosystem development are shown. Later phases will be discussed in Chapter 6.

biomass. This relationship is of exceeding interest because the coupling of decline in dead biomass with increase in living biomass underlies a transfer (which we explore in detail later) of nutrients from dead biomass components to living biomass during the very critical phase of recovery after clear-cutting. It is also interesting to note that the ratio of dead to living biomass changes throughout the 170-yr period, with dead biomass more abundant during the first third of the period.

Study of a number of stands of various ages that have grown up after cutting indicates that there are a number of biogeochemic, hydrologic, and ecologic differences between stands, associated with the descending or ascending portions of the total biomass curve. We propose that the curve of total biomass accumulation is intimately related to the changing dynamics of the ecosystem.

For the purpose of comparative study, we have broken the 170-yr period into two primary parts: the Reorganization Phase from 0 to 15 years, during which the ecosystem has a net loss of biomass, and the Aggradation Phase from 15 to 170 years, when the ecosystem has a net increase in total biomass. Aggradation is a period characterized by accumulation and building of living biomass, forest floor, and dead wood components. It also involves relatively slow species changes, shifts in species dominance, and changes in the structure of the living biomass, as well as the storage of nutrients in biomass.

The term reorganization rather than degradation is used for the 0- to 15-yr period since this period is characterized not only by degradation (in the sense of a net loss of biomass) but also by an enormous and relatively rapid change in the abundance and importance of species (higher plants, microbes, and animals) and changes in biogeochemistry and hydrology. In essence, after disturbance the ecosystem exhibits maximum flux rates and undergoes a rapid but transient reorganization. The net effect is to call into action a variety of processes that limit degradation to a minimum and rapidly return the system to biogeochemic–hydrologic conditions that characterize the Aggradation Phase.

## THE HUBBARD BROOK ECOSYSTEM STUDY

In the next four chapters, our goal is to present data and propose hypotheses that characterize structural, functional, and dynamic aspects of the Aggradation and Reorganization Phases. To do this, we draw heavily on studies done at the Hubbard Brook Experimental Forest. There, in close cooperation with the U.S. Forest Service, we have studied intensively for the last 15 years the hydrology, biogeochemistry, and ecology of six small watershed-ecosystems covered with second-growth forest. One watershed-ecosystem, W6, has served as a reference system, and we have examined in detail species composition, production characteristics, internal nutrient cycling, forest dynamics, and other ecological characters in this ecosystem. One watershed, W2, was experimentally deforested and maintained bare for 3 years before vegetation was allowed to regrow. Another, W4, was clear-cut in strips over a 4-yr period. Behavior of both of these watersheds was monitored during the entire period. Still another watershed, W101, was cut in a commercial forestry operation, and the

**Figure 1-11.** The Hubbard Brook Experimental Forest showing monitored Watersheds 1, 3, 5, and 6, and experimentally manipulated Watersheds 2 (deforested), 4 (strip-cut), and 101 (commercially clear-cut). Note elevational gradient with northern hardwoods giving way to spruce–fir forest at higher elevations and on knobs. (Photograph courtesy of R. S. Pierce, U.S. Forest Service.)

effects were monitored during the recovery period (Figure 1-11). To evaluate the generality of results obtained at Hubbard Brook, we have also studied various aspects of dozens of northern hardwood stands throughout northern New England. In view of the weight placed on data from the Hubbard Brook Ecosystem Study, it seems essential to review study methods and conditions found at Hubbard Brook. General characteristics and biogeochemistry have been treated elsewhere in detail (Likens et al., 1977) and will be briefly summarized here.

## Study Area

The Hubbard Brook Experimental Forest is fairly representative of forests in northern New England growing at intermediate elevations and under relatively oligotrophic conditions (Likens et al., 1967; Bormann et al., 1970; Siccama et al., 1970). The Experimental Forest covers approximately 3000 ha and ranges in altitude from about 200 to 1000-m. The climate is predominantly continental. Precipitation is more or less evenly distributed throughout the year; it averages about 130-cm/yr and is about one-quarter to one-third snow. The bedrock of the area is a medium- to coarse-grained sillimanite-zone gneiss of the Littleton Formation and consists of quartz, plagioclase, and biotite. Generally, bedrock is covered by a shallow layer of glacial till of similar mineralogy (Johnson et al., 1968).

## Soils

Soils are mostly well-drained spodosols (haplorthods) with little clay, a sandy loam texture, and a thick surface organic layer (the forest floor) (Figure 1-7). Soils are acid, pH ≤4.5, and infertile (cf. Pilgrim and Harter, 1977). The principal soil series are Herman (most extensive), Becket, Waubec, Canaan, Berkshire, Peru, Leicester, and Coltin. Surface topography is very rough due to pits and mounds resulting from tree falls and surface boulders. Profiles are very stony, and soil depths are highly variable, averaging about 0.5-m to the till layer. The forest floor tends toward a mor type and depths range from about 3 to 15-cm (Hart et al., 1962; Dominski, 1971; Hoyle, 1973; Likens et al., 1977). Quantitative sampling of the mineral soil presents one of the most difficult sampling problems encountered in the Hubbard Brook Ecosystem Study.

## Hydrology

One of the main concerns about small experimental watersheds is how representative they are of regional hydrology. This is a particularly important question at Hubbard Brook because hydrologic flux is the basis for estimating chemical flux through the ecosystem. Sopper and Lull (1965, 1970) have compared streamflow from the Hubbard Brook experimental watersheds with that of seven other forested watersheds within the northern hardwood region of northern New England. Taking into consideration the relatively small size of the Hubbard Brook watersheds, it was found that streamflow characteristics closely approximate those of the other watersheds and, consequently, that measurements at Hubbard Brook are fairly representative of the region (Likens et al., 1977).

## History of the Hubbard Brook Forest

The lowlands of the Hubbard Brook area (Thornton township) began to be settled by Europeans in 1770. Like the rest of northern New England, the population increased to a peak, approximately 1000 people, about 1830 (Likens, 1972b). Prior to that time, no record exists of cutting at the location of the small watershed-ecosystems we have under intensive study. In 1830 a small sawmill was located at a distance of 5.5-km and at an elevation 400-m below the study sites. It burned and was replaced by another in 1860 (Likens, 1972b). It seems unlikely that much, if any, cutting occurred at that time in the headwater areas of the present Hubbard Brook Experimental Forest, owing to the distance and the rugged terrain and the extensive areas of merchantable timber available at lower elevations. A railroad was completed to nearby North Woodstock in 1883–84, and more intensive logging was possible thereafter (Brown, 1958a,b; Gove, 1968). Some time after 1900, the entire Hubbard Brook watershed was logged. At the study site red spruce was cut around 1909, and all merchantable trees were removed in a heavy cutting completed about 1917 (Bormann et al., 1970). A scattering of large old culls as well as some smaller trees was left. Possibly the forest was selectively cut prior to 1909 (Whittaker et al., 1974). Consequently, the composition of the present-day forest is mostly composed of trees established after the cuts plus a mixture of trees predating the 1909 and 1917 cuts.

## Forest Composition and Elevational Gradients

The present forest at Hubbard Brook is fairly typical of similarly aged second-growth hardwood forests found throughout northern New England at comparable elevations and underlain by similar acid geologic substrate (Bormann et al., 1970; Siccama et al., 1970). Basal area at Hubbard Brook averages about 24-$m^2$/ha, while Barrett (1962) reports 28-$m^2$/ha as fairly representative of second-growth stands in the northeastern United States.

The vegetation of our reference watershed, W6, is fairly typical of present-day forests at intermediate elevations in the White Mountain region of New Hampshire. The overstory is dominated by sugar maple [Importance Value (IV) based on relative basal area, frequency, and density, 33], beech (IV 28), and yellow birch (IV 25), with a mixture of red spruce (IV 4), balsam fir (IV 3), and white birch (IV 4), and the herbaceous layer, 0 to 0.5 m, is dominated by the evergreen fern *Dryopteris spinulosa*, the shrub *Viburnum alnifolium*, and seedlings of sugar maple and beech (Bormann et al., 1970; Siccama et al., 1970; and Forcier, 1973, 1975). *Erythronium americanum*, a vernal photosynthetic, is the second most important herb in terms of aboveground biomass (Muller, 1975). *Oxalis montana, Maianthemum canadense, Dennstaedtia punctilobula, Clintonia borealis, Lycopodium lucidulum, Aster accuminatus*, and *Smilicina racemosa* are important summergreen herbaceous species (Siccama et al., 1970). The vascular flora of the northern hardwood forest at Hubbard Brook contains about 96 species: 14 trees, 11 shrubs, and 71 herbs (Likens, 1973).

The importance values presented above represent average values for W6, which ranges over 245 m of elevation from 546 to 791-m above MSL. Direct gradient analysis (Bormann et al., 1970) indicates that both species distributions and ecosystem structure vary in response to an elevation complex gradient. Tree species are individualistically distributed over the gradient, with principal deciduous tree species gradually dropping out at higher elevations while species common to higher elevation boreal forests gradually increase (Figures 1-12 and 1-13). Structural characteristics of the ecosystem also change. Basal area per hectare, basal area per tree, deciduousness, and canopy height decrease with increasing elevation and increasing nearness to the ridge forming the divide of the watersheds. Conversely, tree density, herbaceous productivity, evergreenness, and species diversity increase with elevation (Figure 1-13). These changes are part of the general transition from northern hardwood forests, which are predominately deciduous, to boreal forests, which are predominately coniferous. The transition occurs at about 760-m in the mountains of New Hampshire.

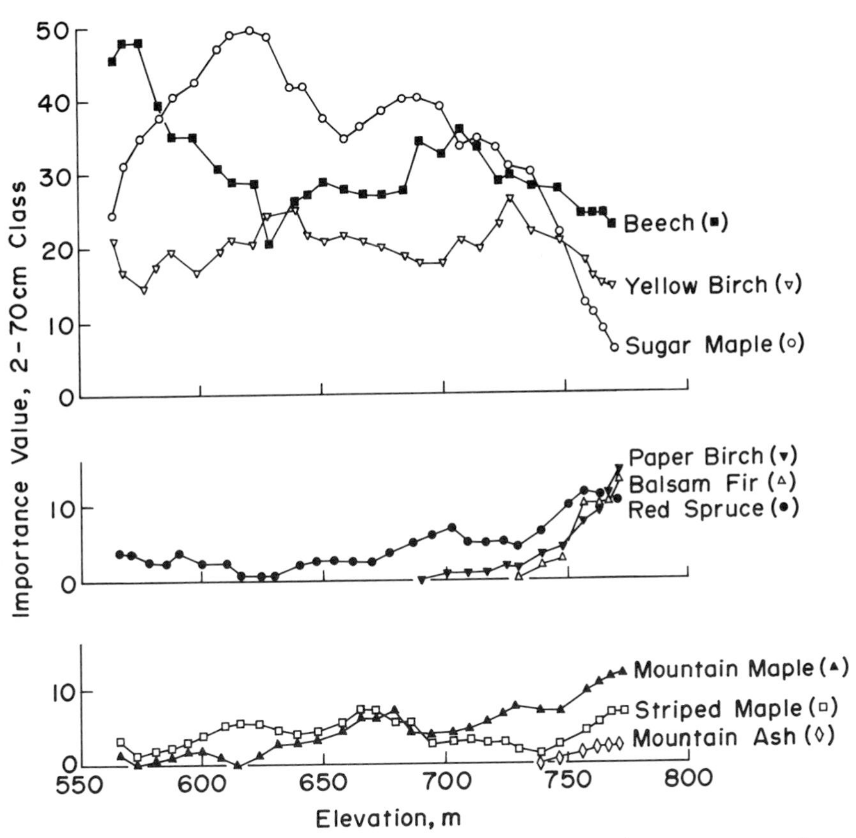


**Figure 1-12.** Running averages of species importance values versus an elevation complex gradient on Watershed 6 are shown for nine species. Points include 30.5-m elevation bands with 7.6-m increments between points. Importance values are based on density, basal area, and frequency in each elevation band. Mountain maple, *Acer spicatum*; Striped maple, *Acer pensylvanicum*; Mountain ash, *Pyrus americana*. Based on 208 randomly stratified plots (from Bormann et al., 1970).

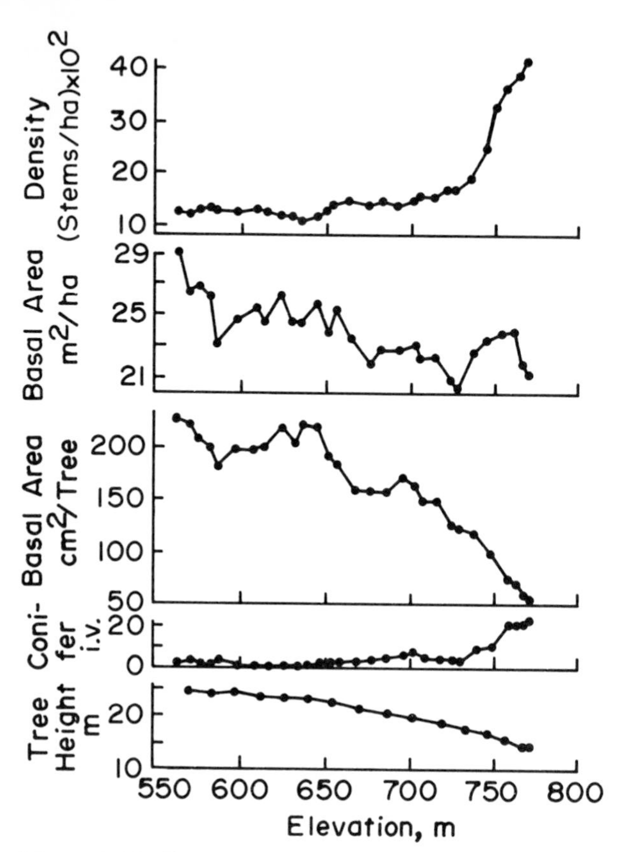


**Figure 1-13.** Running averages of ecosystem characteristics along the elevational gradient on Watershed 6. Points in the average include 30.5-m bands with 7.6-m increments. All values based on stems ≥2-cm dbh on 208 randomly stratified 10 × 10-m plots. Importance values are based on density, basal area, and frequency. The maximum importance value is 100% (from Bormann et al., 1970).

Work by Thomas Ledig and his students has added further to our understanding of spatial variation along the elevation complex gradient at Hubbard Brook. Their study of the photosynthetic response of balsam fir (Fryer et al., 1972) indicated that there was a continuous change in the genetic structure of the population along the gradient such that the temperature at which the photosynthetic maxima occur progressively decreased with altitude roughly in relation to the adiabatic lapse rate (i.e., temperature decline in relation to elevation increase). Other studies indicate that the genetic structure of the sugar maple population is more complex. Sugar maple populations do not differ in their photosynthetic temperature response, and the apparent absence of local "temperature races" may explain the more restricted altitudinal distribution of sugar maple as compared to balsam fir. However, there is a suggestion that sugar maples near the species' upper altitudinal limit have higher rates of photosynthesis than those at midaltitude; perhaps this is an adaptation to compensate for the shorter leafy period at higher altitudes (Fryer and Ledig, 1972). These photosynthetic characteristics affect the patterns of productivity and competition along the elevation gradient.

Ledig's studies indicate that within an ecosystem as small as several hectares there may be significant subspecific populations specially adapted to local conditions, a point that is largely ignored in ordination modeling. Relating this genetic variation to whole-ecosystem parameters is indeed a difficult task. However, the JABOWA Forest Growth Simulator (Botkin et al., 1972a,b) is designed so that genetic variation of the balsam fir type can be indirectly built into the model in such a way that production characteristics will vary in relation to elevation. However, such refinements in ecosystem modeling are years away because, among other things, sufficient genetic information is simply not available.

As a consequence, we do not attempt to deal with genetic variation at an intrapopulation level. Data presented in subsequent chapters are for whole ecosystems that cover a segment of the elevational complex gradient and include a range of genetic variability.

## The Small Watershed Technique as a Method of Biogeochemical Study

In the following discussion of the Aggradation and Reorganization Phases of development after clear-cutting we use biogeochemical data to characterize the phases. These data are of two kinds: input–output budgets in which the ecosystem is considered as a black box and nutrient-cycling analysis where input–output budget data are combined with biogeochemical data internal to the ecosystem to give a more complete picture of nutrient flow within and through the ecosystem. Each of these approaches has its rewards, which are best understood by a brief discussion of several approaches to the study of the biogeochemistry of *humid* terrestrial ecosystems.

Terrestrial ecosystems are open systems, and solid, liquid, and gaseous materials continuously flux through them. Materials are moved through ecosystem boundaries by three major forces or vectors. These are: the meteorologic vector, in which wind or gravity is the motive force; the geologic vector, in which flowing water or gravity is the motive force; and the biologic vector, in which animal power (including man) moves materials from place to place. Meteorologic input to and output from the ecosystem consist of chemicals in airborne organic or inorganic particulate matter, dissolved substances in rain or snow, and gases. Geologic flux includes dissolved and particulate matter transported by surface or subsurface drainage water and mass movement of colluvial materials. Biologic flux results when chemicals gathered by animals or man are deposited in or removed from the ecosystem.

Nutrient and hydrologic parameters of a terrestrial ecosystem (for example, 1 to 100 ha in size) may be studied by considering the ecosystem as a black box in which attention is focused on input and output and little is done to quantify internal relationships. Meteorologic inputs can be fairly readily measured under most circumstances following techniques described in Likens et al. (1977). Quantitative data can be obtained on the amount of water and nutrients added to the ecosystem. If the boundaries of the defined ecosystem are not coincident with those of a small watershed, additional input as well as output vectors must be measured to develop a complete budget (Figure 1-14).

The potential of this nonwatershed approach to biogeochemical study is

THREE LEVELS OF BIOGEOCHEMICAL ANALYSIS OF HUMID TERRESTRIAL ECOSYSTEMS

1. The ecosystem is considered as a black box, and its boundaries are not coincident with those of a small watershed.

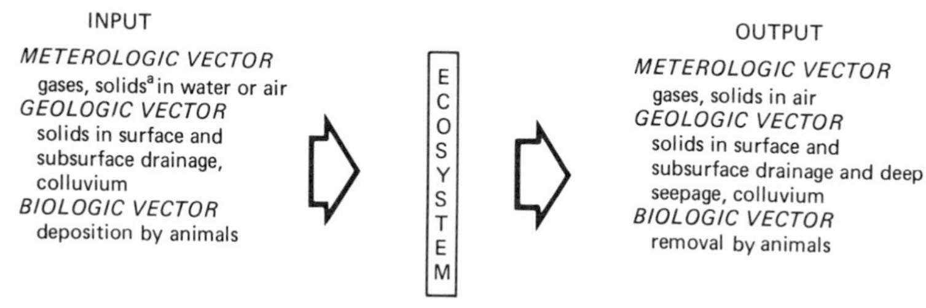


Potential for biogeochemical study

1. Hydrologic and nutrient input in precipitation
2. Partial estimate of geologic output
3. Insight into aspects of ecosystem biogeochemistry

[a]Solids = dissolved chemicals or particulate matter

2. The small-watershed technique where the ecosystem is considered as a black box, but its boundaries are coincident with those of a small watershed underlain by an impermeable substrate. The watershed is in a biologically homogeneous area. It is not necessary to measure input–output vectors shown in italics.

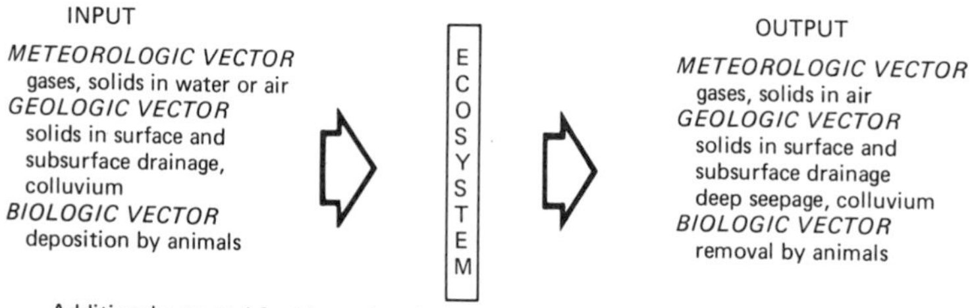


Additional potential for biogeochemical study

1. Complete hydrologic budgets
2. Estimate of evapotranspiration
3. Input–output budgets for sedimentary elements
4. Partial input–output budgets for gaseous elements
5. Net change patterns for individual elements
6. Estimate of minimum weathering rates
7. Establish relationship between streamwater flow rate and streamwater chemistry
8. Insights into annual and seasonal aspects of system biogeochemistry

**Figure 1-14.** Three levels of biogeochemical analysis of humid terrestrial ecosystems.

3. The small-watershed technique coupled with measurement of internal features according to the Hubbard Brook Nutrient Flux and Cycling Model. (After Bormann and Likens 1967, Likens and Bormann 1972.)

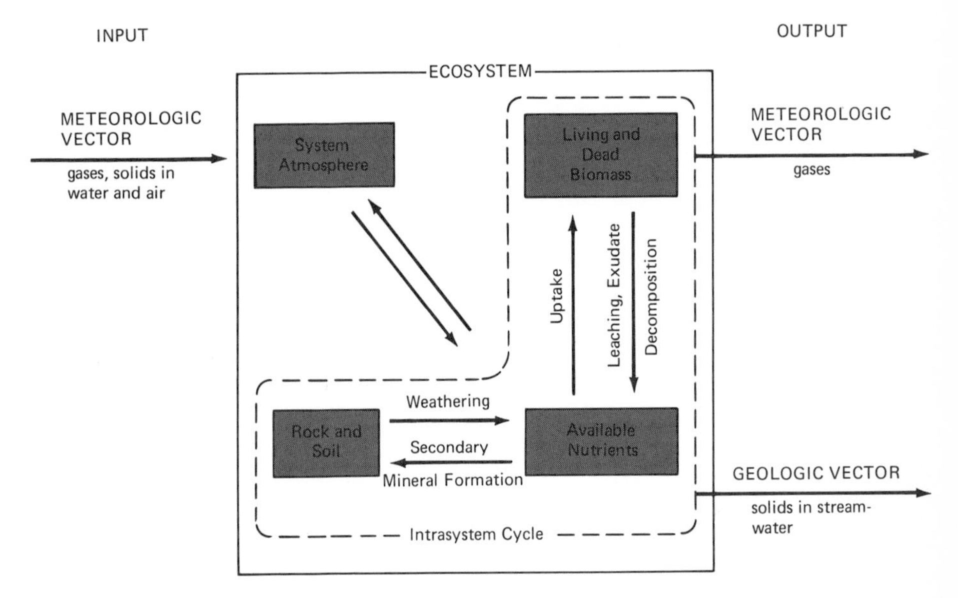


Additional potential for biogeochemical study

1. Annual nutrient uptake by living biomass
2. Nutrient accretion in living and dead biomass
3. Estimates of leaching, exudation, and decomposition
4. Estimates of nutrient input by gaseous uptake, impaction, and nitrogen fixation
5. More complete estimate of weathering

**Figure 1-14** (*Continued from facing page.*)

limited. In many circumstances it is virtually impossible to construct quantitative input–output budgets because of the great difficulty involved in measuring geologic flux into and out of the system. Despite these limitations, the black box approach on nonwatershed ecosystems can yield valuable information on an ecosystem's biogeochemistry. Useful quantitative data can be obtained on the amount of nutrients and water added to the ecosystem by meteorologic input, while measurement of dissolved substances in first-order streams draining the area can give a qualitative idea of one aspect of geologic output as well as some insights into the internal biogeochemistry of the ecosystem.

In some systems that are relatively level and well drained, and where the water table is out of reach, geologic input is minimal, and geologic output might be estimated by lysimeters to yield reasonable input–output budgets.

The small watershed technique adds greatly to the analytical power of the approach just described (Figure 1-14). In this technique, the ecosystem is still considered as a black box, but its boundaries are conceived of as being coincident with those of a small watershed. The technique has several other requirements that add considerably to its usefulness. These are (1) that the watershed be underlain by a watertight substrate so that all liquid water leaving the ecosystem will flow over a weir anchored in the bedrock at the mouth of the watershed, (2) that the watershed be underlain by a uniform geology, so that conclusions regarding weathering rates will not be confused by mixed geologies within the ecosystem, and (3) that the biota of the ecosystem be part of a larger homogeneous biota surrounding the watershed.

At Hubbard Brook all of these conditions are met, and this greatly simplifies the problems of biogeochemical measurement and interpretation (Figure 1-14). Since the lateral boundaries of our ecosystems are coincident with both topographic and phreatic divides surrounding the watersheds, nutrient inputs are thus restricted to meteorologic and biologic vectors since by definition there can be no transfer of geologic inputs between adjacent watersheds. Output is composed of meteorologic, biologic, and geologic output. However, in constructing input–output budgets, major emphasis is on net change within the system and it may be assumed that biologic input balances biologic output because the watershed-ecosystem is part of a much larger more or less homogeneous biologic unit, i.e., second-growth northern hardwood forest. Of course, the biological balance can be upset by such animal activities as timber harvesting, removal of deer by hunters, or possibly differential grazing in recently cutover areas.

The small watershed technique allows, for nongaseous elements, fairly complete input–output budgets. Nongaseous elements are usually not significantly involved in meteorologic output; therefore, budgets may be simply determined from the difference between meteorologic input (dissolved substances and particulate matter in rain and snow) and geologic output (dissolved substances and particulate matter in drainage water).

$$\begin{matrix}\text{Meteorologic} \\ \text{Input}\end{matrix} - \begin{matrix}\text{Geologic} \\ \text{Output}\end{matrix} = \begin{matrix}\text{Net} \\ \text{Change}\end{matrix} \quad (1)$$

The amount and chemical quality of precipitation are measured with a series of rain-gauging stations, and because of the impermeable substrate, the quantity and quality of all drainage water can be accurately measured at the weir. Input–output budgets for elements with gaseous phases are less complete, because this method does not measure either gaseous input or output.

The small watershed black box approach has a considerable potential for biogeochemical study. It allows measurement of a number of important ecosystem parameters. First of all it allows a complete assessment of the ecosystem's hydrologic cycle and an evaluation of the equation:

$$\begin{matrix}\text{Input} \\ \text{(Precipitation)}\end{matrix} = \begin{matrix}\text{Output} \\ \text{(Streamflow)}\end{matrix} + \begin{matrix}\text{Output} \\ \text{(Evapotranspiration).}\end{matrix} \quad (2)$$

This understanding, in its daily, seasonal, and annual dimensions, is basic to the calculation and understanding of the input–output relations of individual chemical elements.

Using input–output hydrologic and nutrient data it is possible to establish the connection between the ecosystem and the larger biogeochemical cycles that influence the ecosystem. Furthermore, man may influence this relationship through air pollution or climate modification. On the other hand, it is through outputs that the terrestrial ecosystem influences the larger cycles as well as the immediately interconnected ecosystems, for example, streams and lakes (Likens and Bormann, 1974b). Man may drastically alter ecosystem outputs by manipulations within the system such as careless forest harvesting or fertilization.

Input–output budgets have considerable value in interpreting ecosystem functions. Consider the meaning of "net change" in Equation (1) for elements with a sedimentary cycle, i.e., without a gaseous phase (Odum, 1971). An algebraically positive net change measured by the small watershed technique means that the ecosystem is accumulating the element from meteorologic input. An algebraically negative net change means that there is a net loss of the element from the ecosystem. That net loss must be accounted for by a decrease of the element somewhere within the ecosystem.

For atmospheric elements, those with a prominent gaseous phase, such as nitrogen, carbon, and sulfur, net change as measured by the small watershed technique has far less interpretive value. This is because gaseous inputs and outputs such as those resulting from biological nitrogen or carbon fixation or denitrification are not measured in Equation (1), which is concerned with measurement of elements associated with movement of water (precipitation and streamflow). Methods for direct measures of gaseous input and output are currently not available.

Input–output budgets also allow an estimate of weathering rates within the ecosystem. In a forested ecosystem, net losses of sedimentary elements like calcium or sodium provide a measure of the rate at which these elements are being released by weathering of substrate minerals within the ecosystem (Johnson et al., 1968). This is a minimum estimate because the small watershed black box technique supplies no information on the accumulation of these elements in various sinks within the ecosystem (e.g., accumulating biomass). Such accumulation may significantly increase the rate of weathering so calculated (Likens et al., 1977).

The small watershed black box technique allows coupling of stream-water flow measurements with measurements of stream-water chemistry and sediment loads. This permits the development of equations relating dissolved substance and particulate matter concentrations to stream-water flux rates, and allows prediction of nutrient output from the hydrologic record alone.

The analytical value of the small watershed technique can be enhanced still further by coupling it with a simultaneous analysis of internal features of the ecosystem. The scientific rewards obtained from this enlargement of scope are very great, but the work load increases enormously. Investigators with limited resources should choose very carefully how they wish to invest their resources in the study of internal features of the ecosystem.

In the earliest phase of the Hubbard Brook Study we designed a nutrient flux and cycling model specifically to exploit the small watershed technique (Bormann and Likens, 1967). One of the major features of the model is that many components and flux rates can be measured directly by well-known techniques. These data, when coupled with input–output data derived from small watershed analysis, allowed for the estimation of still other parameters by the budget balancing method, in which some aspect of a nutrient budget is expressed as an equation with all but one element known, and the unknown is estimated by solution of the equation.

In the Hubbard Brook nutrient flux and cycling model, nutrients are thought of as occurring within the ecosystem in four compartments: atmospheric; living and dead organic matter; available nutrients; and primary and secondary minerals in soil and rock (Figure 1-14). The atmospheric compartment includes all elements in gaseous form both above and below the ground. Available nutrients are ions that are absorbed on clay–humus exchange surfaces or dissolved in the soil solution. The organic compartment includes all nutrients incorporated in living and dead biomass (the dead wood, forest floor, and mineral soil compartments, as discussed above; see Figure 1-5). The soil and rock compartment is composed of unavailable nutrients locked in primary and secondary minerals.

The biogeochemical flux of elements involves an exchange between the various compartments of the ecosystem. Available nutrients and gaseous nutrients may be taken up and assimilated by the vegetation and microorganisms, with some passed on to heterotrophic consumers and made available again through respiration, biological decomposition, exudation, or leaching from living and dead organic matter. Insoluble primary and secondary minerals may be converted to soluble available nutrients through the general process of weathering; soluble nutrients may be redeposited as secondary minerals. Because nutrients with a sedimentary biogeochemical cycle are largely recycled within the boundaries of the ecosystem between the available nutrient, organic matter, and primary and secondary mineral compartments, they tend to form an intrasystem cycle.

Coupling the data obtained from the small watershed technique with measurements of internal features specified by the ecosystem model allows estimates of ecosystem functions not possible when the ecosystem was simply considered as a black box. Perhaps the most important aspect of this relationship is that it allows coupling of energy flow, as measured by productivity data, with nutrient flow as measured by input–output data (Likens and Bormann, 1972b).

At Hubbard Brook we have measured such important productivity features as gross primary productivity, net primary productivity, and living and dead biomass accretion. From these biomass data we have enlarged our understanding of input–output relationships in several ways (Likens et al., 1977). For example, net annual input of gaseous and impacted sulfur may be estimated by subtracting sulfur input in precipitation, sulfur release through weathering, and sulfur accretion in total biomass from sulfur output in stream water. This is a minimum estimate of total sulfur flux because gaseous losses for the ecosystem are not measured. Net nitrogen fixation may be estimated by subtracting nitrogen losses in stream water from nitrogen accretion in total biomass plus input in precipitation (Bormann et al., 1977). Again this is a minimum estimate of gaseous flux since gaseous losses due to denitrification are not measured. Finally, a better estimate of chemical weathering may be made by adding the amount of an element, e.g., calcium or sodium, locked up in biomass accretion to net losses measured by the output–input data obtained by the small watershed technique.

Our nutrient flux and cycling model also provides a framework for quantitative evaluation of the development or degradation of an ecosystem and the effects of changing volumetric dimensions through time (Likens and Bormann, 1972b). This may be expressed in several ways: by (1) alterations in the internal flux rates, (2) changes in the size of compartments, and (3) change in output relationships. These matters will be considered in more detail in subsequent chapters.

## Summary

1. A model of biomass accumulation that portrays the long-term development of the northern hardwood ecosystem is proposed. The model is derived from numerous data collected by scientists of the Hubbard Brook Ecosystem Study and the U.S. Forest Service, and others working in northern New England.
2. The model is constructed to meet certain limits:
    a. It starts with clear-cutting;
    b. it applies to the mountain phase of the northern hardwood ecosystem within a specified geographical and elevational range; and
    c. only minor exogenous disturbance occurs after clear-cutting.
3. The model is based on changes in total ecosystem biomass, i.e., living biomass plus dead biomass in dead wood, the forest floor, and in the mineral soil.

4. The model contains four phases of development: Reorganization, a period of one or two decades immediately after clear-cutting, during which time the ecosystem loses total biomass despite accumulation of living biomass; Aggradation, a period of more than a century when the system accumulates total biomass reaching a peak at the end of the phase; Transition, a variable length of time during which total biomass declines; and the Steady State, when total biomass fluctuates about a mean.
5. The model provides the framework, used in subsequent chapters, to present an integrated picture of ecosystem development that couples biotic and abiotic phenomena.
6. The northern hardwood forest ecosystem at Hubbard Brook is characterized. A nutrient flux and cycling model and the small watershed technique as instruments for ecosystem study are presented and evaluated.

CHAPTER 2

# Energetics, Biomass, Hydrology, and Biogeochemistry of the Aggrading Ecosystem

A logical place to begin our discussion of ecosystem development might be with the Reorganization Phase that immediately follows clear-cutting. However, for a number of reasons that seem to outweigh the risk of a temporary discomfiture to the reader's sense of time, we shall begin our discussion with the Aggradation Phase.

The Aggradation Phase is the one we have most intensively studied. It is characterized by a storage of biomass and nutrients and by maximum biotic regulation over energy, nutrient, and hydrologic flux. Its output relationships are the most predictable of all of the proposed phases. It is our perception that the biogeochemistry and ecology of all the other phases is best understood in terms of departures from the highly predictable parameters of the Aggradation Phase.

Aggradation begins about 15 years after clear-cutting, when the total biomass curve of the ecosystem (Figure 1-10) shifts from net loss to net accumulation. This pattern continues until about Year 170, when total biomass reaches a peak for the ecosystem. The period of aggradation is not one of uniform development or biomass accumulation. The rate of biomass accumulation is at first relatively rapid as accumulation occurs in the living biomass, dead wood, and forest floor compartments (Figure 2-1). Later the rate slows as accumulation is limited primarily to living biomass, which at that time is also accumulating at a lesser rate. Marked changes in forest structure, species composition, and dominance also occur during this 155-yr period (these changes are discussed in Chapter 4). Compared to the Reorganization Phase (from 0 to 15 years after clear-cutting), when there is a net loss in total biomass and fairly rapid and drastic changes in other ecosystem parameters, the Aggradation Phase is

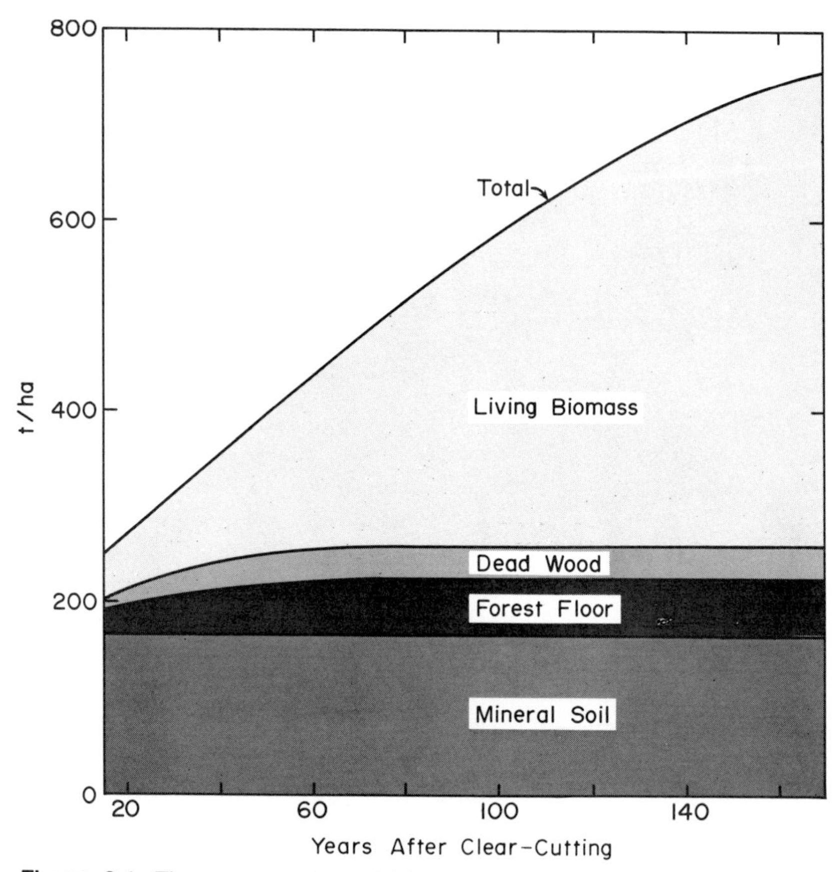


**Figure 2-1.** The accumulation of biomass during the Aggradation Phase of ecosystem development.

one of relative stability. That is, its energetic, hydrologic, and biogeochemic relationships remain fairly constant and predictable. Many of the stability relationships examined in this chapter are for the most part the least subtle ones. More subtle aspects of the long-term relationships are discussed in the chapter that deals with reorganization (Chapter 3).

Forests in the Aggradation Phase currently are found over most of northern New England and are the base from which we have obtained our empirical and experimental data. Since this phase represents a continuum of change, some of the conclusions about ecosystem functions such as weathering, nitrogen fixation, and control of stream-water chemistry based on young aggrading stands (i.e., 50–60 years old) may need modification before they are fully applicable to older stands within the Aggradation Phase. On the other hand, conclusions on hydrology and erosion would seem to apply with little modification to the entire phase. Throughout this section we will draw on information presented in an earlier text, *Biogeochemistry of a Forested Ecosystem* (Likens et al., 1977).

## SOLAR ENERGY FLOW

The stable conditions that characterize the aggrading ecosystem result in large part from its capacity to utilize or control the flow of radiant energy impinging on it. This is done by reflection, heat loss, evapotranspiration, and photosynthetic use. The ecosystem has its greatest effect on energy flow during the growing season (June through September), when 15% of the incoming radiation is reflected, 41% is lost as heat, 42% is used in the evaporation of water (transpiration plus evaporation), and about 2% is fixed in photosynthesis (Gosz et al., 1978).

Solar energy fixed in photosynthesis not only provides most of the energy necessary to drive the biological functions of the ecosystem but is also stored within the ecosystem in the form of the carbon compounds that make up ecosystem structure (Table 2-1). The 55-yr-old forest at Hubbard Brook contains about $7 \times 10^8$ Kcal/ha in living biomass, which accumulates at a net rate of $1.2 \times 10^7$ Kcal/yr. About $1.4 \times 10^9$ Kcal/ha is stored in dead biomass, which accumulates at a net rate of about $1.2 \times 10^6$ Kcal/yr. Living and dead biomass provide the organic structure

**Table 2-1.** Energy Relationships of a 55-Yr-Old Aggrading Forest Ecosystem of the Hubbard Brook Experimental Forest[a]

| | $Kcal \times 10^9/ha$ |
|---|---|
| Energy stored in ecosystem biomass | |
| Living biomass | 0.71 |
| Dead wood | 0.15 |
| Dead biomass (organic matter in soil) | 1.30 |
| Total | 2.16 |
| Incident solar energy | |
| Annual | 10.10 |
| Growing season (June–September) | 4.43 |
| Annual conversion of solar energy | |
| Photosynthesis (GPP) | 0.10 |
| Transpiration | 1.75 |
| Total | 1.85 |
| | Efficiency (%)[b] |
| Photosynthesis alone | |
| Annual | 1.0 |
| Growing season | 2.3 |
| Photosynthesis and transpiration | |
| Annual | 18.3 |
| Growing season | 41.8 |

[a] *Modified from Gosz et al., 1978.*
[b] *Ratio of conversion/incident energy.*

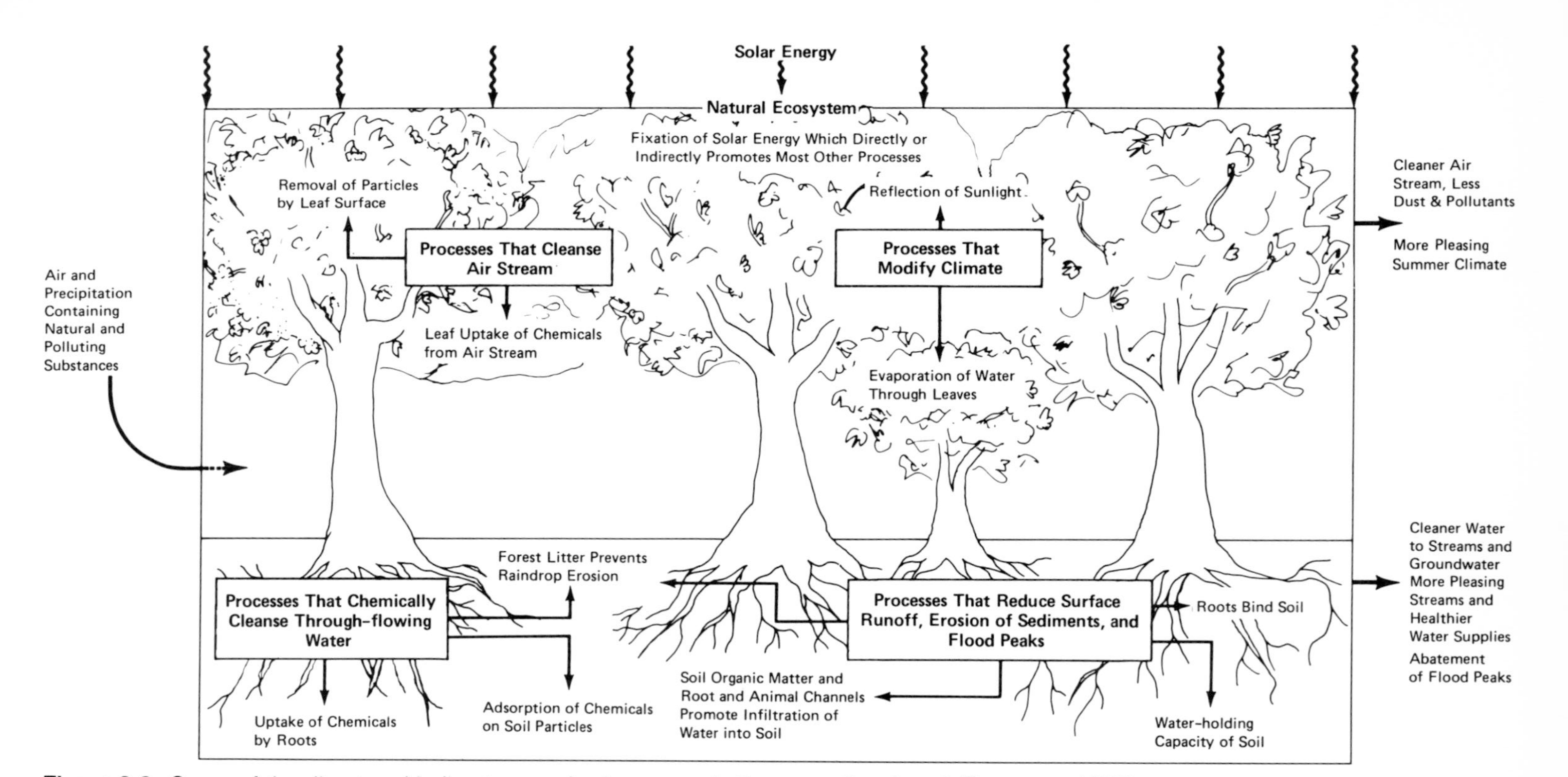


**Figure 2-2.** Some of the direct and indirect uses of solar energy in the aggrading forest (Bormann, 1976).

upon which are based many of the ecosystem's functions—the most notable of which is photosynthesis itself (Figure 2-2). Energy in dead biomass may be considered to be in a "bank" that can be drawn upon in times of stress. This idea will be developed more fully in Chapters 3 and 5.

Photosynthesis is often thought to occur during a growing season defined as the period between the last killing frost in spring and first killing frost in autumn, but in natural ecosystems photosynthetic strategies of component species may extend the energy-fixing season well beyond these points. Muller (1975) has defined three such strategies at Hubbard Brook: summergreen species whose leaf display more or less corresponds to the period between last and first killing frosts, evergreen species that in New England are able to carry on appreciable photosynthesis on either side of the last–first frost (Bourdeau, 1959; Sprugel et al., unpublished), and vernal species that complete much of their life cycle prior to the last killing frost in spring. An analysis of photosynthetic strategies within the Hubbard Brook forest indicates that species with summergreen strategies produce 94% of the aboveground production (Table 2-2). This situation probably holds for most aggrading northern hardwood ecosystems, but the reader should bear in mind that there is a potential for northern hardwood forests to contain a much larger proportion of evergreens.

Finally, we point out that the importance of a particular group of plants in regulating ecosystem function may be greater than their energy-fixing capacity implies. For example, the aboveground activity of the spring herb *Erythronium americanum* is restricted to the period between snowmelt and forest canopy development. Its phenology and production characteristics closely adapt the species to this temporal niche in northern deciduous forests. Although *E. americanum* contributes less than 0.5% of the ecosystem's aboveground production (Table 2-2), it may significantly reduce losses of potassium and nitrogen from the ecosystem during the

**Table 2-2.** Relative Distribution in Percent of Aboveground Production in a 55-Yr-Old Northern Hardwood Forest at Hubbard Brook[a]

| | Photosynthetic Strategy | | | |
|---|---|---|---|---|
| Layer | Evergreen | Vernal | Summergreen | Total |
| Tree | 2.3 | 0 | 94.3 | 96.6 |
| Shrub | <0.1 | 0 | 0.7 | 0.7 |
| Herb | 1.2 | 0.5 | 1.0 | 2.7 |
| Total | 3.5 | 0.5 | 96.0 | 100.0 |

[a] *Distribution is according to photosynthetic strategies and by layers within the forest. Herbaceous layer, 0–0.5 m; shrub layer, 0.5–2.0 m; and tree layer, >2.0 m (after Muller, 1975).*

spring runoff period (Figure 2-3). This temporal pattern suggests that *Erythronium* acts as a short-term sink or "vernal dam," with nutrients incorporated in accumulating biomass during the spring flushing period and released by shoot decomposition during the early summer. Such a mechanism would reduce nutrient losses in spring streamflow, make them available to early summer growth, and thus preserve the nutrient capital of the ecosystem (Muller and Bormann, 1976).

During one year, about $1.3 \times 10^{10}$ Kcal of solar energy are received by each hectare at Hubbard Brook (Table 2-1). About 1% is converted by photosynthesis into gross primary production. Approximately 55% of the GPP is used to sustain green plant respiration, the remaining 45% (NPP) supports all heterotrophic activities and storage in living and dead biomass.

The photosynthetic conversion efficiency of 1%/yr suggests a relatively

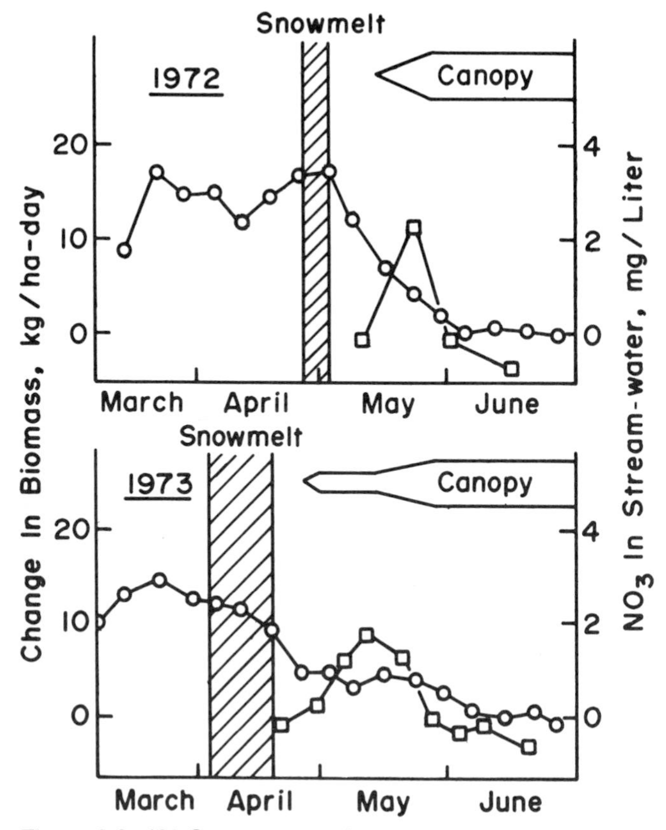


**Figure 2-3.** (○) Stream-water nitrate concentrations and (□) biomass increase in *Erythronium americanum* in 1972 and 1973. Biomass increase was calculated as the rate of biomass change between any two consecutive harvest dates. Period of snowmelt and closure of the overstory canopy are shown (Muller and Bormann, 1976).

modest capacity of the ecosystem to capture and use solar energy. However, the utilization of solar energy in photosynthesis supports life and growth of the green plants and makes it possible for the ecosystem to utilize a large amount of solar energy in the evaporation of water by the process of transpiration. As a result, water that would leave the ecosystem as liquid water in streamflow is converted to a gas and leaves as water vapor. The $1.7 \times 10^9$ Kcal/ha-yr of solar energy used in transpiration (the calculations of energy used in transpiration will be given in Chapter 3) greatly enhances the capacity of the ecosystem to resist external destabilizing forces to which it is continually subjected, such as wind, water, and gravity.

The use of energy in transpiration represents about 13% of all incident energy received by the ecosystem during 1 year. When energy used in transpiration is added to that of photosynthesis, annual efficiency is 14%. However, if we calculate efficiency on the basis of incident energy in June through September, the months when almost all production and transpiration occurs, efficiency is about 37%.

Solar energy channeled into photosynthesis and transpiration may be thought of as sustaining active and passive ecosystem processes which greatly affect the overall biogeochemistry of the ecosystem. Transpiration, carbon assimilation, nitrogen fixation, and water and nutrient uptake may be thought of as active processes that require a continuous expenditure of biologically fixed or biologically mediated solar energy. Whereas passive processes may be thought of as nonbiological (physical or chemical) processes that require little immediate expenditure of biologically fixed or biologically mediated solar energy, such as reflection of sunlight, impaction of aerosols, filtration, and exchange processes in the soil. These processes remove chemicals or particles from the streams of air and water moving through the ecosystem (Figure 2-2). The relationship between active and passive processes is complex. For example, soil exchange processes may be viewed as passive, yet biologically derived exchange sites on humus play an important role in the process. Impaction, the process of removal of aerosols and dust from the airstream, presents another interesting case. Although this is a physical process, the biological structure and composition of the ecosystem exercises considerable control over the intercepting surface and thus affects the rate of impaction. The architecture of the major impaction sites, i.e., the vegetation, changes both on a seasonal basis with deciduousness and on a longer-term basis with ecosystem development.

Through regulation of the flow of solar energy and the use of solar energy to support both passive and active processes, the aggrading ecosystem gains substantial control over internal microclimate as well as biogeochemistry. Regulation of internal humidity and temperature regimes in summertime is achieved by the reflective capacity of the upper canopy and by transpiration, which evaporates approximately $3 \times 10^6$ liters of water/ha-yr at Hubbard Brook. The highly buffered microclimate within the system plays a major role in regulating the decomposition regime within the soil, and in this way is probably closely related to the highly predictable biogeochemistry of the aggrading ecosystem (Likens et al., 1977).

## BIOMASS: DEVELOPMENT OF REGULATION AND INERTIA

The amount and distribution of living and dead biomass are among the most important structural features of northern hardwood ecosystems and are closely and functionally linked to their stability.

Analysis of the second-growth forest at Hubbard Brook (Table 2-3) provides data roughly equivalent to a 55-yr-old stand in our hypothetical sequence (Figure 1-10). Total biomass is about 42,000 g/m². Of this, about 16,000 g/m² is living plant biomass distributed about 82% aboveground and 18% belowground. Eighty percent of the fine roots (<3-mm) are concentrated in the upper 30 cm of the soil (Safford, 1974). Although the forest has a leaf area index of 5.8, only 315 g (2.4%) of the aboveground biomass is in leaves (Whittaker et al., 1974). The bulk (61%) of the total biomass is dead. However, inspection of Figure 2-1 suggests that living biomass predominates from about 80 years after clear-cutting to the end of the Aggradation Phase. It should be noted, however, that accumulation of living biomass is primarily the accumulation of wood in the interior of living trees.

Concentrations of both living and dead biomass vary markedly

**Table 2-3.** Living and Dead Biomass in a 55-Yr-Old Aggrading Forest Ecosystem at Hubbard Brook[a]

| Category of Biomass | Vertical Distribution (m) | Biomass (g/m²) | |
|---|---|---|---|
| | | Living | Dead |
| Above soil surface | | | |
| Living plant biomass | 0.00–20.0 | 13,281[b] | |
| Dead wood | 0.00–20.0 | | 440 |
| Below soil surface | | | |
| Living roots | 0.00–0.5 | 2,826[b] | |
| Dead wood | 0.00–0.5 | | 2,900 |
| Forest floor | 0.00–0.09 | | 4,800 |
| Mineral soil | 0.09–0.45 | | 17,300 |
| Total | | 16,107 | 25,440 |

[a]*From Dominski, 1971; Whittaker et al., 1974; Gosz et al., 1976; Melillo, 1977.*
[b]*Does not include herbs and shrubs, but these are <1%.*

throughout the vertical dimension of the ecosystem. The concentration of biomass per unit volume in the upper 45-cm of the soil is about 90 times greater than average concentrations aboveground. This distribution reflects the functional requirements of distributing leaves through a large volume of air to facilitate energy interception and gas exchange and of concentrating roots in a relatively small volume of soil containing available nutrients and water.

## Dead Biomass and the Forest Floor

About 60% of the organic matter below the soil surface (Table 2-3) is incorporated in the mineral soil, but concentrations are quite variable spatially owing to pit and mound topography and the generally rocky nature of the profile. Most (90%) of the organic matter in the upper 45 cm of the soil is dead and is largely the product of microbiological alterations of plant material. The mineral soil is generally covered with a well-developed forest floor which averages about 8.6 cm (Dominski, 1971; Melillo, 1977) in depth and contains about 4800 g/m$^2$ of organic matter (Figure 1-7). Organic matter in the forest floor and in the B-horizon of the mineral soil are not the same. The organic matter of the forest floor is considered more labile and nutrient-rich, while most of the organic matter in the deeper horizons is thought to be more refractory and less subject to rapid change (see Chapter 1).

Dead biomass in the soil has very strong effects on both hydrologic and nutrient-cycling characteristics of the ecosystem. Organic matter reduces bulk density, increases water-holding and cation-exchange capacity, and serves as a reserve store of plant nutrients (Hoyle, 1973). Because of the low bulk density of the forest floor (0.2 kg/liter; Dominski, 1971), these soils, in common with most other forest soils (Lull and Reinhart, 1972), have enormous infiltration capacities (ca. 76-cm/hr; Trimble et al., 1958), as well as excellent percolation rates. Because of the relative lack of clay in our northern hardwood soils, humus also serves as the principal cation-exchange site within the soil profile, and the highest concentrations of exchangeable cations are found within the forest floor (Hoyle, 1973; Pilgrim and Harter, 1977).

The growth of the forest floor during the Aggradation Phase, from 26 t/ha at 15 years after clear-cutting to 57 t/ha at 80 years after cutting (Covington, 1976), indicates that both available water-storage and cation-exchange capacity grow as well. We estimate that between Years 15 and 60 water retention capacity (mostly available water storage; Trimble and Lull, 1956) grows by 0.7-cm and that the cation-exchange capacity of the forest floor increases by about 100%. Since water and nutrients are vital to production, it would appear that the productive capacity of the aggrading ecosystem increases during the first third of the Aggradation Phase coincident with the growth of the forest floor.

## DETRITAL–GRAZING CYCLES

The detrital cycle is overwhelmingly dominant in terms of energy and nutrient flux within the aggrading northern hardwood ecosystem. In an average year we estimate that about 26% of the net primary productivity in a 55-yr-old aggrading forest is stored within the ecosystem in the form of a net gain in living biomass, mostly wood and bark (Figure 2-4). Seventy-four percent of NPP plus a small addition of dissolved organic matter in precipitation represents that proportion of the current crop of energy available to support the sum of all heterotrophic activities. Less than 1% of this energy is consumed by grazing animals in an average year. The principal herbivores, in approximate order of importance, are chipmunks, mice, foliage-eating insects, birds, deer, and hares (Gosz et al., 1978; Hanson, 1977). The remaining energy is added to the dead

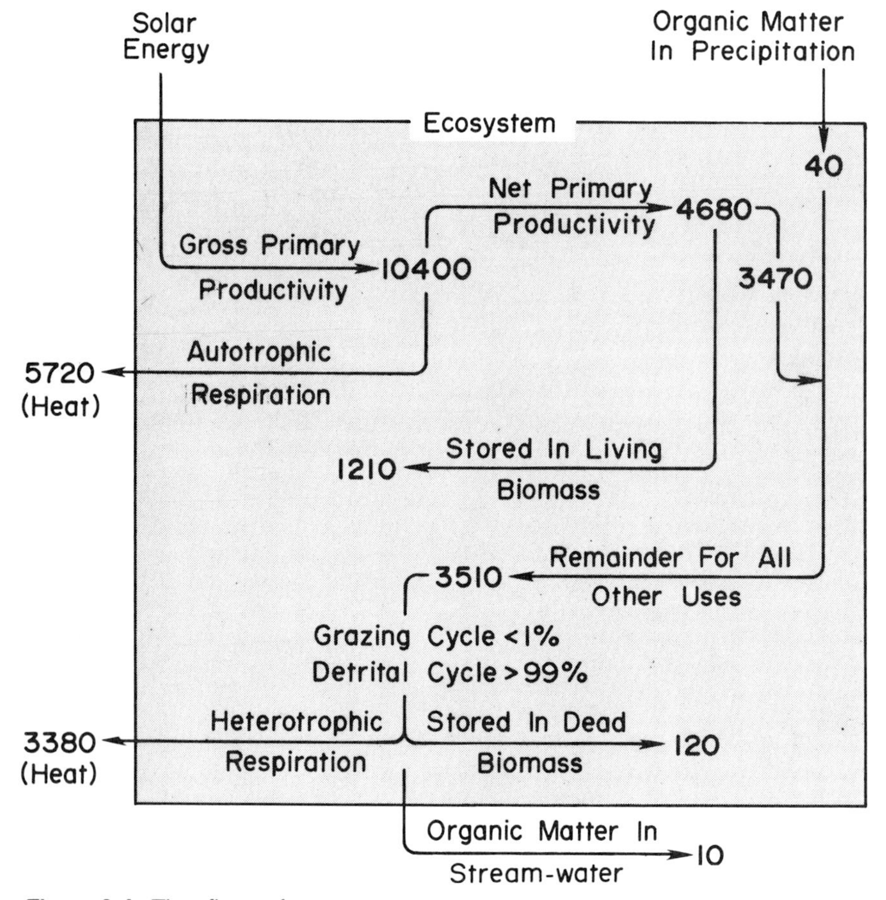


**Figure 2-4.** The flow of energy in carbon compounds through a 55-yr-old aggrading forest at Hubbard Brook; numbers give energy values in kilocalories per square meter (after Gosz et al., 1978).

wood, forest floor, and mineral soil compartments in various forms of litter, leachates, and exudates. There it is subject to utilization in the detrital cycle (Figure 2-4). In an average year, an amount of energy equivalent to about 95% of that added to the detrital cycle (but not necessarily the same energy) is released as heat due to heterotrophic respiration, while about 5% is added to net dead biomass accumulation.

Energy consumed in all heterotrophic respiration, grazing and detrital, is coupled to the biogeochemical process of mineralization or the release of ions previously locked in organic compounds. In effect, these ions are added to the available nutrient compartment in our nutrient flux and cycling model (Figure 1-14).

We have not identified the principal saprobic plants (microflora) or invertebrate detritivores at Hubbard Brook; however, the major groups have been reported for acid mor soils elsewhere in the Northeast (Eaton and Chandler, 1942; Lutz and Chandler, 1946). Fungi are thought to be the principal agents of decomposition since the relative importance of bacteria declines markedly in acid soil. Major invertebrate detritivores are probably similar to those reported for mor soils in New York State where, in terms of numbers, the following groups predominated: Arachnida (mites, false scorpions, and spiders), Collembola (springtails), Coleoptera and Diptera (beetles and flies), Hymenoptera (ants), Symphyla, and Annelida (earthworms). Earthworms, millipedes, and isopods were absent in the New York State sample, but all have been found in small numbers at Hubbard Brook (R. Holmes, personal communication).

The complex relationship and relative importance of the microflora and detritivores in decomposition remains to be determined at Hubbard Brook. However, it would appear that as soon as soft tissues such as leaves or bud scales fall to the forest floor they are subject to a coordinated attack (Lutz and Chandler, 1946; Jackson and Raw, 1966; Schaller, 1968; Brady, 1974). Fungi and bacteria initiate the action but are soon joined by springtails, bark lice, and various larvae which eat or tear holes in the tissue (fenestration), opening it to more rapid microbial attack. Larger larvae and mites bring about further perforation and skeletonization. Large amounts of feces, or frass, are produced, which may be consumed again by other fauna. The activities of the soil fauna and microflora are thus closely linked. Chewing, ingestion, and digestion by fauna not only result in decomposition of the organic matter but simultaneously create surface and moisture conditions more favorable to microbial action both within the faunal gut and in the resultant frass. It seems likely that the detritivores obtain their principal energy supplies from the easily decomposable substances within the litter such as sugars, starches, and simple and crude proteins. Exoenzymes of fungi and bacteria not only attack these easily decomposable substances but are largely responsible for the decomposition of the more resistant compounds, such as hemicellulose, cellulose, and lignins, which compose the bulk of the leafy and wood litter.

The rate of detrital decomposition by the microflora and the detritivores is regulated by many factors including the dimensions and physical and chemical structure of the substrate. In general, boles of trees, particularly large ones, decay slowly, while leaves decay fairly rapidly. Litterbag studies at Hubbard Brook (Gosz et al., 1973) indicate that yellow birch leaves decompose more rapidly than sugar maple leaves, which in turn decay more rapidly than beech; we estimate that 5, 9, and 11 years, respectively, are required for 95% of the original dry weight of the above species to disappear after a leaf falls to the forest floor.

Soil invertebrates also play a role in the physical transport of decomposing organic matter. However, owing to the relatively small populations of earthworms and other larger invertebrates at Hubbard Brook, there is relatively little mixing activity within the forest floor or with the underlying mineral soil. The relatively happy circumstance of a fairly clear distinction between the forest floor and the mineral soil made it possible for us to carry out the study of the decline and growth of the forest floor discussed in Chapter 1.

The microflora and detritivores are themselves food sources for parasites and predators, including carnivores such as mites, centipedes, salamanders, and shrews, as well as microfloral feeders. To some degree, mites and springtails regulate fungal populations by feeding on hyphae (Mitchell and Parkinson, 1976), and protozoa may exercise some control over bacterial populations (Lutz and Chandler, 1946).

Lack of knowledge of the details of saprobic decomposition represents one of the major gaps in our knowledge at Hubbard Brook, yet this is of key importance in understanding the ecosystem's biogeochemistry, and particularly changes through time. For example, on the basis of input–output budgets (Figure 1-14) we know that when an aggrading ecosystem is clear-cut the output of nitrate may increase severalfold. This is due primarily to a marked increase in the populations and activities of autotrophic nitrifying bacteria (Smith et al., 1968) and other heterotrophic nitrifying organisms (J. Duggin, personal communication) within the soil. What we do not know is how the rest of the saprobic plant–detritivore system responds. Do the new conditions result in marked expansion or contraction of various detritivore and microfloral populations? A comparative study of changes in saprobic populations in the forest floor of undisturbed and recently clear-cut forests would seem to be an informative way of attacking this problem.

The division of energy flow between the detrital and grazing food webs is not constant from year to year, and in some years there is a dramatic rise in herbivory, principally due to rapid increases in populations of leaf-eating insects. High rates of defoliation occurred at Hubbard Brook in the years 1969–71 when a species of defoliating caterpillar, the saddled prominent (*Heterocampa guttivita* Walker), was in outbreak phase. During the peak year of defoliation, about 44% of the total leaf tissue was consumed, while in local areas patches of forest were totally stripped of leaves (Holmes and Sturges, 1975; Gosz et al., 1978). Leaves of yellow birch were preferred by the larvae of the defoliating insects (R. Holmes, personal communication). Although these insects consume large quantities of leaf tissue, the amount of energy they assimilate is relatively small, about 14% of the total contained in the ingested tissue. The remainder (86%) is added to the detrital cycle in the form of frass.

The potential of these organisms to affect system biogeochemistry goes far beyond their energy utilization, because their grazing activities strike directly at the ecosystem's primary source of energy fixation and hydrologic and energy regulation, which is leaf tissue. Not only is there a strong potential to reduce gross primary productivity temporarily and hence diminish the amount of energy available to support all organisms within the ecosystem (Figure 2-4), but by reducing leaf area it seems likely that the ecosystem would undergo proportionate reductions in transpiration and nutrient uptake by green plants. According to Hibbert's review (1967) of various forest-thinning experiments, streamflow is increased approximately in direct proportion to the reduction in leaf area. In undisturbed hardwood forests of North Carolina, it has been observed that defoliation by the fall cankerworm (*Alosophila pometaria*) is associated with concentrations of nitrate in stream water that are about five times higher than those measured in nondefoliated watersheds (Swank and Douglass, 1975). Increased nitrate loss is thought to result from the influence of defoliation on the forest floor, for example, from higher soil temperatures or lower nutrient uptake by plants. At Hubbard Brook, no significant increase in nitrate concentration in stream water was observed during 1970, the year of peak insect defoliation.

The following effects emphasize the large destabilizing potential that severe outbreaks of leaf-eating insects may have on an ecosystem: (1) conversion of living leaf tissue to $CO_2$, insect biomass, and frass; (2) increased output of stream water, the chief vehicle for removing nutrients; (3) decreased uptake of nutrients by green plants, perhaps leading to increased nutrient concentrations in soil and stream water; (4) more radiant energy flow to the forest floor; (5) increased transfer of living biomass to dead organic matter (Figure 1-5); and (6) possible alterations in microbiologic activities in the soil leading to increased nitrification (implications to be discussed later). Production may be temporarily lowered, and the system may become more open in terms of stream-water and nutrient loss. Not only does defoliation affect major aspects of the system's biogeochemistry, but the detrital cycle also must be affected. During heavy outbreaks of defoliators, leaves are consumed in the canopy, and we wonder what effect this has on the detritivore populations. For example, certain populations of soil invertebrates carrying out fenestration and skeletonization rely heavily on newly fallen leaf litter (Schaller, 1968).

It is exceedingly difficult to evaluate the long-term role of defoliating insects in regulating the biogeochemistry of natural ecosystems. On the one hand, insects may constitute a major destabilizing force causing numerous rate changes and an increased loss of nutrients from the ecosystem. On the other hand, because of differential grazing and other internal ecosystem responses, they may act to channel the flow of radiant energy, nutrients, and water to better-adapted individuals and species and thus function something like a cybernetic regulator of primary production (Mattson and Addy, 1975). At Hubbard Brook these defoliating organisms apparently operate within bounds that allow the aggrading ecosystem to continue its accumulation of living biomass at the fairly predictable pace reported by Whittaker et al. (1974).

Holmes and Sturges (1975) and Gosz et al. (1978) report a behavior pattern of predatory animal species at Hubbard Brook that may illustrate another animal-based biogeochemical regulatory mechanism. They point out that many of the major predators are almost entirely opportunistic in their diet preferences and that these animals cannot be identified with either the grazing or detrital cycles. In most years, however, it seems likely that major predators such as birds, shrews, salamanders, and rodents and invertebrates such as centipedes, beetles, and spiders gain their major energy inputs by feeding on a variety of detritivores and their invertebrate predators. They also consume defoliating insects, and during years when they are common many of these predators have the capacity to substantially shift their diet to the newly available food source.

The intensive study of birds by Holmes and Sturges (1975) provides data for one group at Hubbard Brook. In general, birds consume only a small part, about 0.2%, of the annual net primary productivity occurring aboveground and probably have no major direct effect on ecosystem structure or overall energy flow rates. During average years, when most energy is flowing to the detrital cycle, the summer diet of the bird community consists largely of adult Dipterans whose larvae are detritivores in the forest floor and in streams, adult Coleopterans whose larvae prey on soil-dwelling organisms, and adult Hymenopterans which are themselves secondary and tertiary consumers. In most years, then, the bird community is closely linked by energy flow with the forest floor. However, Holmes and Sturges (1975) have shown that during the years when *Heterocampa* was in outbreak phase, many birds concentrated on this new and abundant source of food. During the years of outbreak, 1969 to 1971, and for one year afterward, the total number of individuals in breeding bird populations increased progressively, presumably because of a more abundant and easily available food supply. Studies of animal behavior during outbreaks of gypsy moths suggest that other predatory populations such as mice (*Peromyscus*) and beetles (*Calasoma*) behave in a fashion similar to the bird population at Hubbard Brook (Campbell, 1975).

These behavior patterns suggest that these vertebrate and invertebrate populations of opportunistic predators, although playing a minor role in energy flow patterns, may perform important biogeochemical regulatory functions in both the detrital and grazing cycles. They must play some role in regulating the process of mineralization through effects on detritivore populations and the processes of transpiration, nutrient cycling, and primary production through effects on populations of insect defoliators. Holmes (personal communication) believes that populations of defoliating insects at Hubbard Brook are kept at low numbers by two factors: (1) defense systems of the plants themselves, including the low food quality of plant tissues and (2) predators. When plant defenses are lowered because of stress or other factors, the insects consume more plant tissue, reproduce and grow faster to a point where they are uncontrollable by the predators, and thus enter an outbreak phase. Campbell (1975, personal communication) suggests that once an outbreak begins other factors such as parsitoids, disease, and starvation are much more important in bringing about its termination. It is most interesting that these diverse predators are firmly attached to the detrital cycle, the most predictable aspect of energy flow within the ecosystem, and in a sense are facultative with regard to the less predictable grazing cycle.

## BIOTIC REGULATION OF BIOGEOCHEMICAL FLUX

### Regulation of the Hydrologic Cycle

Personnel of the U.S. Forest Service have monitored and maintained accurate records of precipitation and streamflow at Hubbard Brook since 1956 (Likens et al., 1977; Figure 2-5). On the average the area receives 130-cm/yr of precipitation, of which 70% is rain and 30% is snow. In aggrading forests, about 62% runs off as liquid water in stream channels, while the remainder is lost as vapor through evapotranspiration.

In any particular year, monthly inputs of precipitation may show random extremes, but for the longer term the monthly pattern is quite regular (Figure 2-5). In contrast to seasonal uniformity of precipitation, monthly streamflow is widely variable and in fact defines four seasons (Figure 2-5). Streamflow during the "growing season," approximately June through September, is very low. Most of the annual streamflow, 54%, occurs during the spring snowmelt period, March through May, with 30% in April alone. A second minor peak in streamflow occurs from mid-October through mid-December, after leaf fall but before heavy development of the snowpack; whereas streamflow is diminished during the winter, January through mid-March, as cold winter conditions prevail and precipitation accumulates as a snowpack.

The living and dead biomass of the Aggradation Phase exercise considerable and fairly predictable control over the hydrologic cycle for the ecosystem. Control is expressed largely by effects on the storage of water within the ecosystem and on the pathways by which water moves through and out of the ecosystem.

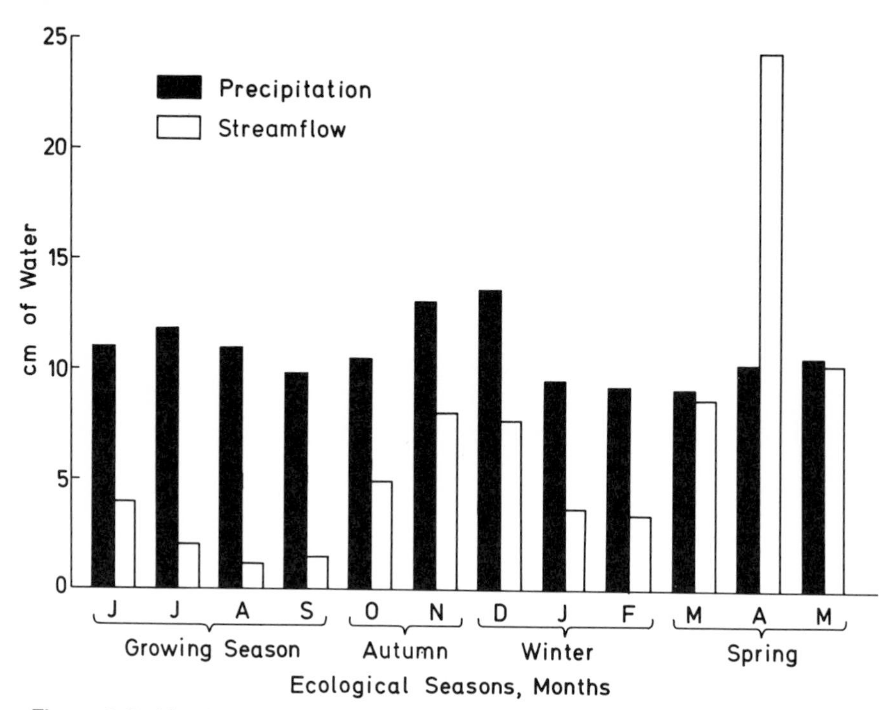


**Figure 2-5.** Monthly average distribution of precipitation and streamflow in centimeters of water. Ecological seasons are the growing season; autumn; winter, accumulating snow; and spring snowmelt and runoff period (after Likens et al., 1977).

The living biomass of the aggrading ecosystem affects the streamflow characteristics during all seasons (Figure 2-5). During the autumn, increased streamflow is correlated with the cessation of transpiration. In the hard winter, and in spring snowmelt periods, timing of streamflow is affected by the shade of standing leafless trees, but this is of minor importance. "Growing season" effects, however, are important and concern not only timing but amount of streamflow. In this, the biological process of transpiration plays a central role.

Hydrologic data from June through September, when the forest is in full leaf, show that, while precipitation ranges from 32 to 50-cm, streamflow averages only 5-cm. As is well known (Kittredge, 1948; Colman, 1953; Hibbert, 1967), during the growing season evapotranspiration (mostly transpiration) has first call on water stored in soil, thus evacuating storage space. Only after storage space is refilled by precipitation does streamflow occur.

During the Aggradation Phase, no destruction of the forest canopy is visualized and thus it seems probable that the hydrologic conditions just described are generally applicable to the entire period, although minor variations due to variable proportions of evergreen trees might occur (Swank and Douglass, 1974).

The effect of transpiration is so strong that when 18 years of data on annual evapotranspiration (which includes evaporation throughout the year as well as transpiration during the growing season) and annual streamflow are graphed against annual precipitation (Figure 2-6), evapotranspiration remains relatively constant despite major fluctuations in precipitation, while streamflow is highly correlated with amount of precipitation. This indicates that the aggrading ecosystem has a remarkably constant transpirational demand over a wide range of environmental variables.

When precipitation crosses the boundary of an ecosystem, it has both potential energy and potential solvent power which, as it passes through and out of the ecosystem, could be utilized in the erosion and transportation of particulate matter and in the removal of dissolved substances. The vegetation of the aggrading ecosystem acts to minimize these degrading activities by the utilization of solar energy in transpiration. Thus streamflow is reduced, and a variety of other mechanisms allow

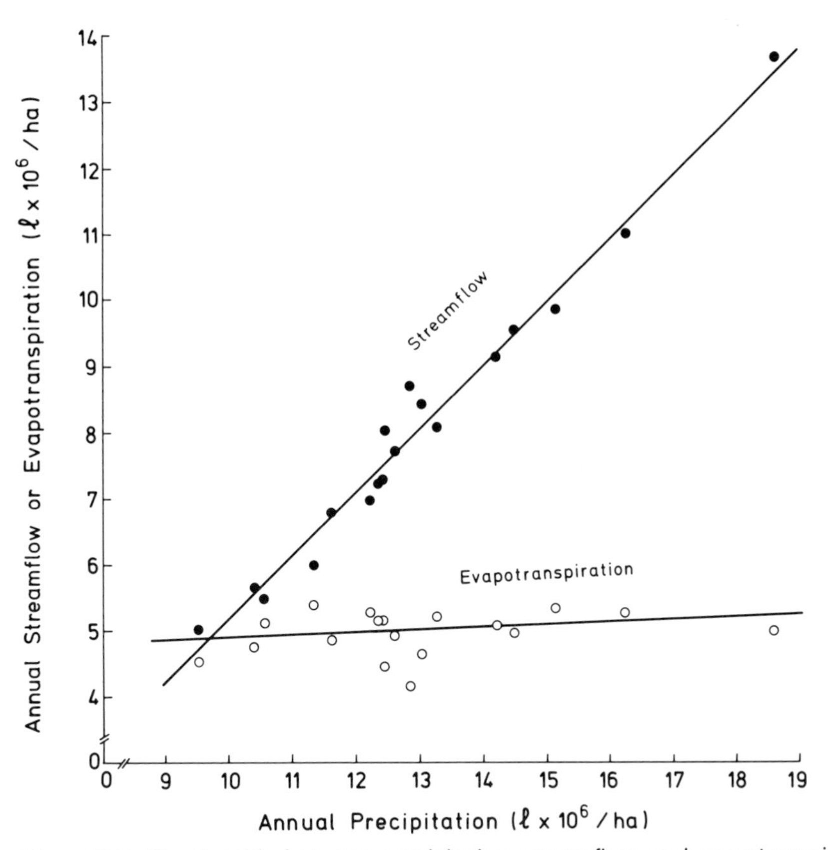


**Figure 2-6.** Relationship between precipitation, streamflow, and evapotranspiration for the Hubbard Brook Experimental Forest during 1956–74 (Likens et al., 1977).

the ecosystem to absorb the kinetic energy of flowing water while yielding a minimum of eroded material.

Physiologists often portray transpiration as a necessary evil, or at the very most as useful in cooling leaf tissue on hot days. From the broader perspective of the ecosystem, in humid systems transpiration is a major regulatory process exercising important control over a variety of ecosystem responses. Solar energy, impinging on the forest canopy, is utilized to increase evaporation over what it would be if no vegetation were present. This transfer of liquid water to vapor has these actual or potential ecosystem effects:

1. Transpiration markedly reduces stream-water output during the growing season. We estimated an average annual reduction of stream-water output of 25% at Hubbard Brook due to transpiration (Pierce et al., 1970).
2. This reduction acts as a nutrient conservation mechanism because reduced streamflow means less nutrients carried out of the ecosystem.
3. Transpiration not only reduces streamflow, but also diminishes summertime-peak discharge rates during storms (Reinhart et al., 1963; Pierce et al., 1970; Hornbeck, 1973b; Bormann et al., 1974). This regulation has important effects on the process of erosion, since erosion is closely related to storm peaks. Obviously, during exceedingly heavy and/or prolonged periods of rain, transpiration would have relatively little effect on flood peaks.
4. Transpiration "powers" a large part of the circulation of nutrients within the ecosystem. Some nutrients are drawn to root surfaces in the mass flow of water immediately after rainfall and some are lifted to the canopy by transpirational "pull," and in contrast to streamflow, which carries nutrients out of the system, transpiration may be viewed as a "distillation" process in which nutrients are left behind in the leaves where they are eventually recirculated by leaf drop, resorption, or leaching.
5. Finally, the complicated positive-feedback relationship between transpiration and availability of soil water plays an important role in regulating primary production of the ecosystem through well-known effects of available water on photosynthesis (Bormann, 1953; Kramer and Kozlowski, 1960) and, indirectly, in regulating decomposition within the soil. Transpiration also may be of some importance as a soil-aerating mechanism by removing soil water after rainfall and, in effect, opening air passages into the soil. This effect is well known for wet sites (Dansereau, 1957).

Biomass not only exerts control over the hydrologic cycle by transpiration but in several additional ways. Living leaves and forest litter intercept and disperse the energy of falling drops of water. The heavy concentrations of dead biomass in the $A_0$-horizon and in the mineral soil greatly augment both retention and detention water storage in the soil and create excellent conditions for infiltration and percolation of water. This latter factor determines the route of liquid water passing through the ecosystem, because it permits almost unlimited infiltration. Overland flow is negligible in forested ecosystems at Hubbard Brook (Pierce, 1967); as a consequence, water moves to the stream by lateral flow within the soil. This relationship exists even during the winter, because organic layers coupled with snow accumulation insulate the soil, and normally there is no soil freezing during the winter (Hart et al., 1962). This routing of water has important implications for both ecosystem biogeochemistry and erosion. Finally, interception and evaporation of both rain and snow by dormant and nondormant northern hardwood forests at Hubbard Brook reduce the amount of precipitation reaching the ground by about 12% (Leonard, 1961).

## Control of Particulate Matter Export

In terms of overall ecosystem stability, soil erosion is perhaps the most important potential destabilizing force for terrestrial ecosystems. Chronic erosion of exposed mineral soil tends to remove the finer inorganic fractions and organic matter. These fractions are rich sources for available nutrients and also contribute to the moisture-holding capacity of the soil; hence erosion has the potential to seriously diminish ecosystem production (Brady, 1974). In its most severe form, erosion can remove a significant proportion of both organic and inorganic exchange surfaces within the ecosystem. Not only will this have immediate drastic effects on the productivity of the ecosystem, but it may have substantially long-term effects. The rebuilding of exchange surfaces to pre-erosion levels by the production of clay minerals through weathering and accumulation of soil organic matter and the reacquisition of lost nutrients may take several centuries. In essence, the developed ecosystem may be forced back to a more primitive level of development with lower production and less control over its immediate environment. Loss of control may for a time act as a positive feedback and delay the rate at which the ecosystem develops after serious erosion. It is interesting to note that this most basic and important perturbation to ecosystem stability, erosion, long known to agriculturalists and foresters, hardly enters the current ecological debate on stability of ecosystems. In fact, the word erosion is not even found in the subject index in a number of recently published ecology texts.

Aggrading northern hardwood ecosystems in the Hubbard Brook area have low rates of erosion due to geological circumstances (Hunt, 1967; Lull and Reinhart, 1972), but the biological fraction of the ecosystem also plays an important regulatory role. Rates of erosion of particulate matter for second-growth forests are among the very lowest recorded for humid ecosystems. Losses average $3.3 \pm 1.3$ t/km$^2$-yr, despite the fact that these forests occur on rather steep slopes, 12 to 13° and are subject to an average annual precipitation of 130-cm (Figure 2-5). This output is small compared to other forests and represents an extremely low rate when compared to potential outputs of thousands of metric tons per year from seriously disturbed sites (Table 2-4).

At Hubbard Brook we have found, as have others elsewhere, that particulate matter losses are directly but nonlinearly related to the discharge rate of the stream (Figure 2-7). The shape of the concentration–flow rate curve is determined by the erodibility of the ecosystem and the fact that the capacity of running water to do work increases roughly as the square of its velocity (Bormann et al., 1969). Erodibility is a complex function of the ecosystem, and disturbance of the ecosystem can shift the curve (Figure 2-7) to the left, making low flow rates more effective in the erosion of particulate matter.

**Table 2-4.** Particulate Matter Output (Sediment Yield) from Forested Ecosystems, Small River Watersheds with Mixed Cultural Conditions, and Small Man-Manipulated Ecosystems in the Northeastern United States[a]

| Location | Watershed Size ($km^2$) | Sediment Yield ($t/km^2$-yr) |
|---|---|---|
| *Forested ecosystems* | | |
| Mature forest ecosystem, Hubbard Brook (8-yr average) | 0.12 | 3.3 |
| Protected second-growth forest, mountains, West Virginia | 0.39 | <3.2 |
| Wooded watershed, Somerset, Kentucky | 2.2 | 5.3 |
| *Selected small rivers* | | |
| Tributary basins | | |
| Susquehanna River, 88% forest cover | ? | 12 |
| Potomac River, 88% forest cover | ? | 12 |
| Streams draining New England upland in New Jersey | ? | 3.5–35 |
| Scantic River, Connecticut | 253 | 26 |
| George Creek, Maryland (rural and wooded) | 188 | 72 |
| Tributary basins | | |
| Susquehanna River, 25% forest cover | ? | 85 |
| Potomac River, 15% forest cover | ? | 160 |
| *Small man-manipulated ecosystems* | | |
| Deforested ecosystem, Hubbard Brook | | |
| (4-yr Average) | 0.16 | 19 |
| (Maximum annual) | 0.16 | 38 |
| Careless clear-cut, mountains, West Virginia | 0.30 | 302 |
| Clear-cut, followed by farm and pasture, mountains, North Carolina (maximum yield) | 0.09 | 3,600 |
| Construction site, Baltimore, Maryland | 0.006 | 49,000 |

[a] *Modified from Bormann et al., 1974.*

An analysis of particulate matter output and flow rates for the aggrading forest at Hubbard Brook shows the importance of the occasional storm peak in the removal of particulate matter (Bormann et al., 1974). During a 50-month period, 77% of the total stream water was discharged at rates between 0 and 30 liters/sec, but only 14% of the particulate matter was removed. At discharge rates of >85 liters/sec, 3.7% of the water was discharged, while 52% of the total particulate matter was lost. Sixteen percent of the particulate matter was eroded by one flow ranging from 310 to 340 liters/sec. This flow carried only 0.2% of the total streamflow and comprised less than 1 hour's time during the entire 50-month period (Figure 2-8). Clearly, the annual variability in particulate matter output from the aggrading forest is due primarily to the occurrence of large storms.

The biological component of the aggrading ecosystem regulates particulate matter erosion by its effect on both discharge rates and erodibility. As already noted, transpiration during the growing season tends to reduce the size of storm peaks. This is of special note since erodibility shows seasonal variability and is greatest during the growing season (Bormann et al., 1969, 1974).

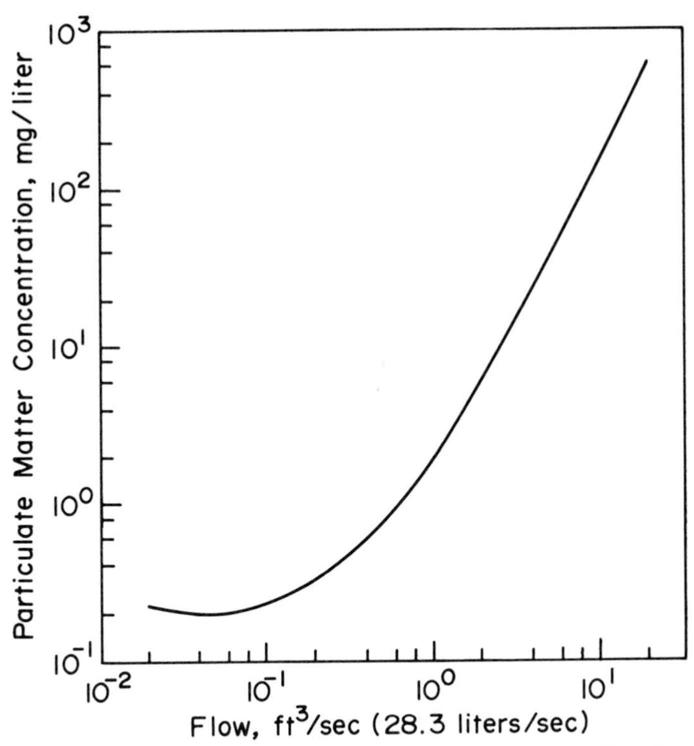


**Figure 2-7.** Composite curve for second-growth aggrading forests at Hubbard Brook (Bormann et al., 1969) showing the relationship between particulate matter concentration in stream water and flow rate.

**Figure 2-8.** (A) The weir at Watershed 4 showing a flow of 0.02 $ft^3$/sec (0.6 liters/sec). Average annual flow of W4 is 0.32 $ft^3$/sec (9.2 liters/sec).

**Figure 2-8.** (B) Photograph taken 24 October 1959, when the weir on W4 was under construction and during the highest flow recorded at Hubbard Brook during its 24-yr history. The peak discharge measured on Watershed 3, 1.2 times larger than W4, was 68.3 $ft^3$/sec (1934 liters/sec). According to a log-Pearson plot, the storm of 24 October 1959 has a recurrence interval of about once every 50 years. One storm of this magnitude would probably remove an amount of particulate matter equivalent to many years of removal at flow rates not marked by unusually high storm flow. The scale in the foreground of both photos is about the same. (Photograph B courtesy of R. S. Pierce, U.S. Forest Service.)

Erodibility, a measure of the ease with which earth materials give way (Leopold et al., 1964), is a function of both geologic substrate and biomass. Whether or not the kinetic energy of moving water within the ecosystem is strongly coupled to the process of erosion and transportation of particulate matter or is dissipated in other ways is often determined by the biotic condition of the ecosystem. That the biota and its organic debris exert considerable control over this phenomenon has been shown experimentally and will be discussed in more detail in Chapter 3.

Biotic control of erodibility in the aggrading forest is achieved by: (1) organic matter enhancement of infiltration and percolation, which converts surface flow to subsurface flow (surface flow does the work of particulate matter erosion); (2) the binding together of soil and organic debris in drainage channels and along stream banks by roots; (3) protection (interception) of mineral soil from direct action of falling raindrops by canopy and forest floor layers; (4) organic-debris dams in stream channels which tend to regulate streamflow and retain particulate matter within the ecosystems; and (5) hardwood leaves, which form a protective shield against erosion of exposed stream banks (Bormann et al., 1969; Fisher, 1970).

## Regulation of Streamflow Chemistry

An important characteristic of the aggrading hardwood forest ecosystem is its ability to regulate the chemistry of drainage water. The chemistry of water entering the ecosystem as rain or snow is altered in highly predictable ways as it passes through the system and issues forth as stream water.

The measurement of precipitation chemistry at Hubbard Brook represents the longest comprehensive record of precipitation chemistry in the United States (Likens et al., 1977). Precipitation is acid, and sulfate and hydrogen ions dominate (Table 2-5). The chemical composition of precipitation is highly variable on a weekly basis, but, in general, relatively lower concentrations occur with snowfall (Likens et al., 1977).

Several long-term trends, apparently related to air pollution, have emerged from these studies. Nitrate concentrations have approximately doubled since 1955. The input (concentration times amount of precipitation) of nitrate and hydrogen ions has increased by 2.3 and 1.4 times, respectively, during the last decade (Likens et al., 1977).

What happens to the chemistry of precipitation as it passes through the ecosystem? Because of the dense forest canopy, the heavy concentration of organic matter in the forest floor, and the high infiltration capacity of the soil, almost all precipitation entering Hubbard Brook watersheds comes into intimate contact with the living and dead biomass and the mineral soil of the ecosystem. During the growing season, first contact of the incoming water (precipitation) is made with the forest canopy. The chemistry of the resulting throughfall and stemflow is markedly changed,

**Table 2-5.** Volume-Weighted Concentrations of Dissolved Substances in Precipitation and Stream Water from Aggrading Forest Ecosystems at Hubbard Brook with Values as Means for the Period 1963–74[a]

| Substance | Precipitation (mg/liter) | Streamflow (mg/liter) | Streamflow/Precipitation Ratio |
|---|---|---|---|
| $Ca^{2+}$ | 0.16 | 1.65 | 10.3 |
| $Mg^{2+}$ | 0.04 | 0.38 | 9.5 |
| $K^+$ | 0.07 | 0.23 | 3.3 |
| $Na^+$ | 0.12 | 0.87 | 7.2 |
| $Al^{3+}$ | [b] | 0.24 | — |
| $NH_4^+$ | 0.22 | 0.04 | 0.18 |
| $H^+$ | 0.073 | 0.012 | 0.16 |
| $SO_4^{2-}$ | 2.9 | 6.3 | 2.2 |
| $NO_3^-$ | 1.47 | 2.01 | 1.4 |
| $Cl^-$ | 0.47 | 0.55 | 1.2 |
| $PO_4^{3-}$ | 0.008 | 0.002 | 0.25 |
| $HCO_3^-$ | ~0.006 | 0.92 | 153 |
| Dissolved silica | [b] | 4.5 | — |
| Dissolved organic carbon | 2.4 | 1.0 | 0.42 |

[a] *From Likens et al., 1977.*
[b] *Not determined, trace quantities.*

in part owing to cation-exchange reactions between precipitation and foliage. Some ions like sulfate and potassium are greatly enriched in the throughfall and stemflow, whereas others like hydrogen ion are depleted (Eaton et al., 1973; Wood and Bormann, 1974; Likens et al., 1977).

Lack of soil frost during the winter (Hart et al., 1962) and the high infiltration and percolation capacity of the soil mass guarantees that during all seasons almost all water moving through the ecosystem to the stream channel comes into intimate contact with the exchange surfaces within the soil. The chemistry of water that follows this route is remarkably altered; concentrations of some ions rise more than 10-fold, whereas those of others drop more than 6-fold (Table 2-5). Increases in concentration result in part from evapotranspiration losses which reduce the amount of liquid water leaving the ecosystem. Theoretically, this could increase concentrations by 1.6 times, but decreased concentrations and the size of some increases (Table 2-5) show that factors in addition to evapotranspiration are operational. In fact, various complex biogeochemical mechanisms are involved in determining the chemical composition of water as it passes through the ecosystem. Some of these mechanisms will be discussed below.

In spite of the numerous and marked differences in the chemistry of water entering the ecosystem as precipitation and that leaving as streamflow, stream-water concentrations are highly regulated by the aggrading forested ecosystem. Concentrations of the various chemicals in stream water are either nearly constant or highly predictable (Johnson et al., 1969; Likens et al., 1977). This is remarkable since streamflow varies by three or four orders of magnitude during the year. The maintenance of this relatively stable stream-water chemistry is related to the intimate contact of the water with exchange surfaces as it travels toward the stream channel and to an abundance of exchangeable ions in relation to ions lost in transient waters (Likens et al., 1977). Factors controlling ultimate stream-water concentrations, however, are very complex and involve regulation of and interaction with numerous aspects of both the hydrology and biogeochemistry of the ecosystem. There is little doubt, however, that activities or conditions associated with living and dead biomass play a major role in regulating this stable and predictable behavior of the aggrading ecosystem.

## Predictability of Stream-Water Chemistry

Johnson et al. (1969) developed a dilution–concentration model for Hubbard Brook which explained transient changes in the chemistry of stream water by means of mixing processes (i.e., mixing of precipitation with soil water). The rate of the mixing process is proportional to the discharge of stream water from the watershed. The model showed three patterns of behavior for chemicals moving through the ecosystem:

1. those that were diluted with high discharge—$Na^+$ and dissolved silica;
2. those that were more concentrated at high discharge—$H^+$, $Al^{3+}$, and $NO_3^-$; and
3. those whose concentrations changed relatively little in relation to stream discharge—$Mg^{2+}$, $Ca^{2+}$, $K^+$, $SO_4^{2-}$, and $Cl^-$. During the growing season, concentrations of $NO_3^-$ and $K^+$ in stream water were markedly reduced.

Before proceeding further, attention should be directed to the close relationship that exists between annual gross output of dissolved substances (kilograms per hectare) in stream water and the annual output of stream water in area centimeters (Figure 2-9). This relationship represents one of the most predictable aspects of the aggrading ecosystem. At first glance it might appear to be at odds with the dilution–concentration behavior of the various elements at different discharge rates described above. However, changes in nutrient concentrations in stream water, although significant for some ions, are relatively small (generally less than three times minimum values), while changes in stream discharge range over four orders of magnitude. This, plus the fact that the bulk of the streamflow (77%) occurs at flow rates between 0 and 30 liters/sec (Bormann et al., 1974), provides the strong predictable relationship between annual gross output of dissolved substances and total annual streamflow.

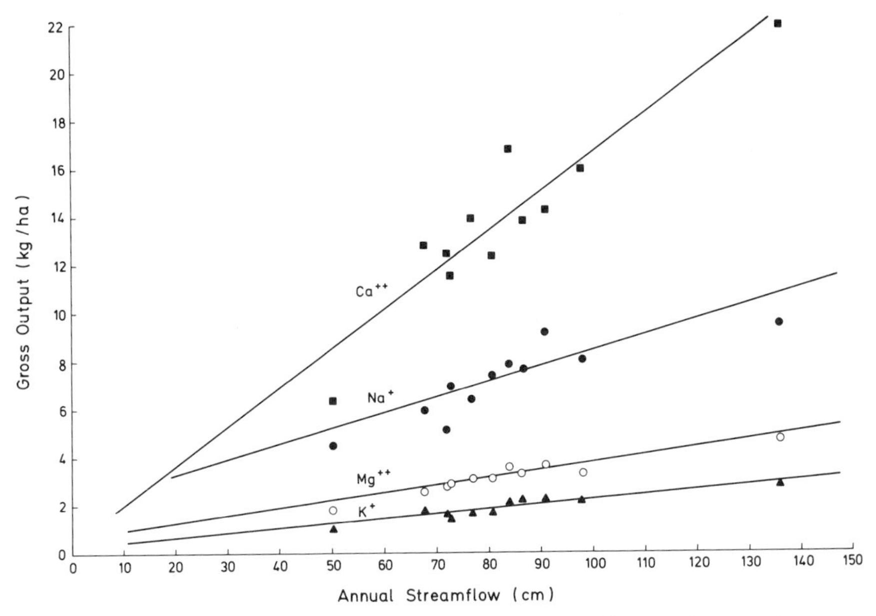


**Figure 2-9.** Relationship between total annual streamflow and annual gross output of calcium, sodium, magnesium, and potassium (in kilograms per hectare) during 1963–74 for forested Watersheds 1–6 at the Hubbard Brook Experimental Forest (Likens et al., 1977).

## Nutrient Input–Output Budgets

Eleven years of data from six small watershed-ecosystems at Hubbard Brook indicate that nutrient losses from the aggrading forested ecosystem are relatively small (Table 2-6). Also, it may be observed that, for some elements, meteorologic inputs are very important in offsetting geological outputs. In fact, the input–output budgets suggest that absolute amounts of hydrogen ions, phosphorus, and nitrogen are increasing within the aggrading ecosystem at Hubbard Brook; i.e., amounts in bulk precipitation exceed amounts exported in stream water. Although the average data (Table 2-6) indicate that chloride is increasing within the ecosystem in the long term, data for annual bulk precipitation are highly variable, and thus inputs are not statistically greater than outputs (Likens et al., 1977).

Input–output budgets take no account of gaseous exchanges for elements like sulfur and nitrogen. It can be seen from Table 2-6 that bulk precipitation does not account for all losses of sulfur in stream water. Since sulfur is only a minor constituent of bedrock and till at Hubbard Brook, it thus seems probable that net losses cannot be made up by weathering within the ecosystem. In fact this is the case. Weathering releases only about 0.8 kg of S/ha-yr (Likens et al., 1977). So to provide the sulfur necessary for annual biomass (living and dead) accretion (Table

**Table 2-6.** Average Annual Input and Output (Losses) of Nutrients for a 55-Yr-Old Aggrading Forested Ecosystem at Hubbard Brook with Values in Kilograms per Hectare per Year During the Period 1963-74[a]

| Element | Meteorologic Input | | Geologic Output–streamflow[b] | | | Net Loss |
|---|---|---|---|---|---|---|
| | Bulk Precipitation | Net Gas and Aerosol | Dissolved Substances[b] | Organic Particulate Matter[b] | Inorganic Particulate Matter[b] | |
| Aluminum | [c] | [c] | 2.0 | [c] | 1.4 | 3.4 |
| Silicon | [c] | [c] | 17.6 | 0.1 | 6.1 | 23.8 |
| Calcium | 2.2 | [d] | 13.7 | 0.06 | 0.17 | 11.7 |
| Magnesium | 0.6 | [d] | 3.1 | 0.05 | 0.14 | 2.7 |
| Potassium | 0.9 | [d] | 1.9 | 0.01 | 0.51 | 1.5 |
| Sodium | 1.6 | [d] | 7.2 | <0.01 | 0.25 | 5.9 |
| Iron | [c] | [c] | [c] | 0.01 | 0.63 | 0.64 |
| Hydrogen ion | 0.96 | [c] | 0.10 | — | — | −0.86 |
| Phosphorus | 0.04 | [d] | 0.01 | <0.01 | 0.01 | −0.02 |
| Nitrogen | 6.5 | 14.2[e] | 3.9 | 0.11 | [c] | −16.7 |
| Sulfur | 12.7 | 6.1[e] | 17.6 | 0.02 | 0.01 | −1.2 |
| Chloride | 6.2 | [d] | 4.6 | [c] | [c] | −1.6 |

[a] *After Likens et al., 1977.*
[b] *Dissolved substances and organic particulate matter constitute net output of elements that occur or have occurred in ionic form within the ecosystem; inorganic particulate matter is largely composed of unweathered primary or secondary minerals.*
[c] *Nearly zero.*
[d] *Relatively small.*
[e] *Partly due to biological activity within the ecosystem, i.e., nitrogen fixation, gaseous absorption, and impaction on plant surface.*

2-7) and net steam-water losses, an additional 6.1 kg/ha-yr is required from some external source. We suggest that this source of sulfur is meteorological input related to biological activity in the canopy of the forest (see Figure 1-14); that is, sulfur is removed from the throughflowing airstream by vegetation. Such sulfur is not effectively measured by bulk precipitation collectors. Sulfur as $SO_2$ may be taken up directly from the airstream by living vegetation (Olsen, 1956; Hoeft et al., 1972), or $SO_2$ and sulfur-rich aerosol particles may be accumulated on foliage surfaces by dry deposition. Support for this latter conclusion is indicated by our study of throughfall and stemflow chemistry within the aggrading forest (Eaton et al., 1973). The amount of sulfur moving to the forest floor in throughfall and stemflow during the growing season is several times greater than that found in precipitation impinging on the forest canopy and apparently cannot be accounted for by rapid cycling within the ecosystem.

Nitrogen in bulk precipitation exceeds stream-water losses by 2.5 kg/ha-yr (Table 2-6). Some 16.7 kg of N/ha are stored each year in living and dead biomass (Table 2-7). Thus, as was the case for sulfur, the nutrient budget for nitrogen indicates that even after we account for input in bulk precipitation, there is a net gain for nitrogen within the aggrading ecosystem of 14.2 kg/ha-yr that must be provided from some source. The most likely source is nitrogen fixation, since the inorganic weathering substrate contains negligible quantities of this element. Therefore, we assume that 14.2 kg/ha-yr is a minimum average estimate of nitrogen

**Table 2-7.** Average Annual Net Accretion of Nutrients in Biomass of the Aggrading Forest Ecosystem 55 Years after Clear-Cutting[a]

| | Nutrients in Biomass (kg/ha-yr) | | | |
|---|---|---|---|---|
| Nutrient | Living Biomass[b] | Dead Wood | Dead Biomass in Forest Floor[c] | Total Biomass Accretion |
| Calcium | 8.1 | 0 | 1.4 | 9.5 |
| Magnesium | 0.7 | 0 | 0.2 | 0.9 |
| Sodium | 0.15 | 0 | 0.02 | 0.2 |
| Potassium | 5.8 | 0 | 0.3 | 6.1 |
| Phosphorus | 2.3 | 0 | 0.5 | 2.8 |
| Sulfur | 1.2 | 0 | 0.8 | 2.0 |
| Nitrogen | 9.0 | 0 | 7.7 | 16.7 |
| Iron | 1.5? | 0 | 1.2 | 2.7? |

[a] *Total biomass accretion is the sum of net nutrient accumulation in three biomass subcompartments: living biomass, dead wood, and forest floor.*

[b] *Living biomass is taken as an average of two pentads, 1955–60 and 1960–65 (Whittaker et al., 1974); nutrient concentrations from Likens and Bormann (1970).*

[c] *Forest floor biomass from Covington (1976); nutrient concentrations from Dominski (1971), Gosz et al. (1976), and Melillo (1977).*

fixation in this 55-yr-old aggrading ecosystem (Bormann et al., 1977). The estimate is minimum because denitrification, which would remove N from the ecosystem in a gaseous form, would not be detected by our budget measurements. The estimate is average because the input–output budgets are averaged over 10 years (Likens et al., 1977) and because the biomass accretion value is an average of two pentad periods, 1956–60 and 1961–65. In a more complete budget, nitrogen losses due to denitrification would have to be added to losses in stream water, and hence estimated nitrogen fixation could be larger than that predicted. We also have no estimate of the input of nitrogenous gases or aerosols. If these are present in significant amounts, the input due to nitrogen fixation would be reduced accordingly.

Of the two types of nitrogen fixation, symbiotic and free living, the latter is more likely since among the higher plants common in the aggrading forest (Bormann et al., 1970; Siccama et al., 1970) none is known to have symbiotic nitrogen fixation. Free-living nitrogen fixation in mixed hardwood forests has been reported by Todd (1971), and this topic has been studied at Hubbard Brook by J. Roskoski (1977). Results based on the acetylene-reduction technique (Hardy et al., 1968) indicate that decaying wood, lying on the surface of the ground, is an active site of free-living nitrogen fixation by as yet unidentified microbes. Roskoski's preliminary estimates indicate that small amounts of N (~1 kg of N/yr) are added to the ecosystem by fixation in woody litter within the ecosystem.

All other elements (Table 2-6) show net losses from the ecosystem. Since the ecosystem is aggrading, or growing, and these elements have no prominent gaseous phase, we can conclude that these net losses from the intrasystem nutrient cycle must be made up by the weathering of primary and secondary minerals in bedrock and till. In our budgetary scheme for nutrient flux and cycling, bedrock and till undergoing weathering are considered to be within the ecosystem's boundaries.

## NUTRIENT RESERVOIRS WITHIN THE AGGRADING ECOSYSTEM

Nutrients stored in reservoirs or sinks within the ecosystem provide inertia and ameliorate change in functional characteristics of the ecosystem. On the other hand, reservoir size or rate of filling may limit or enhance the rate of ecosystem development. It is apparent that nutrient losses from the aggrading ecosystem (gross output of dissolved substances and organic particulate matter; Table 2-6), plus immobilization of nutrients (internal sinks), must be balanced by nutrient additions (sources) if the ecosystem is to continue to grow and develop. As we have already shown in the previous discussion of nitrogen, the relationship (output + sinks = sources), provides a useful analytical tool for probing various aspects of the biogeochemistry of the aggrading northern hardwood ecosystem.

Our nutrient flux and cycling model (Figure 1-14) indicates that there are three potential nutrient sinks within the aggrading ecosystem: (1) in biomass, (2) in available nutrients, and (3) in secondary minerals forming within the ecosystem. For any element, the sum of the net accretion of nutrients in these compartments would equal the amount immobilized within the ecosystem during some period of time (Table 2-7).

The aggrading ecosystem by our definition is accumulating biomass in three subcompartments—living biomass, dead wood, and forest floor (Figure 2-1); and we have estimates of nutrient accretion in each subcompartment (Table 2-7).

We have no complete measure of exchangeable or available nutrients for the entire soil profile at Hubbard Brook. However, Hoyle (1973) indicates that, for soils similar to Hubbard Brook, a large proportion of the available nutrients are located in the forest floor. Since our estimates of elements in the forest floor includes atoms incorporated in the molecular structure of biomass and exchangeable nutrients (Covington, 1976), a large part of any net change in available nutrients is already incorporated in the nutrient data associated with the total biomass accretion values in Table 2-7.

The net accumulation of elements in secondary minerals in the soil is very difficult to quantify directly on a short-term basis. The deeper soils at Hubbard Brook are filled with rocks and boulders of various sizes, and the soil depth is highly variable (Figure 1-7). Such heterogeneity makes quantitative sampling extremely difficult. However, a comparison of elements removed from the ecosystem [net output of dissolved substances and particulate organic matter (Table 2-6) plus that immobilized in accumulating biomass (Table 2-7)] with their abundance in the weathering substrate (Table 18 of Likens et al., 1977) provides some indication about secondary mineral formation. Output of inorganic particulate matter is not included because of its unweathered nature (Table 2-6). Evaluation of the removal abundance ratio shows that aluminum has the lowest ratio at Hubbard Brook, indicating that in proportion to its abundance in rock and till it is the least removed. The order is Al < Si < Fe < K < Mg < Na < Ca.

The low ratios for aluminum, iron, and silicon, coupled with the fact that aluminum, iron, and silicon oxides are major secondary minerals in podzolic soils at Hubbard Brook (Johnson et al., 1968), are suggestive that secondary mineral formation is occurring and that mineral soil profiles are still developing.

## SOURCES OF NUTRIENTS FOR THE AGGRADING ECOSYSTEM

Undisturbed watershed-ecosystems have two major sources of nutrients: (1) meteorologic input of dissolved, particulate, and gaseous chemicals from outside the ecosystem's boundaries and (2) release by weathering of nutrients from primary and secondary minerals stored within the ecosystem's boundaries (Figure 1-13). Biological activity within the ecosystem may affect the rate of both meteorologic input and weathering. For example, the architecture of the forest canopy may determine the rate of removal of airborne aerosols moving through the ecosystem, while the release of nutrients by weathering is influenced by a variety of intrasystem biotic and abiotic factors.

### Weathering

When the various aspects of flux, cycling, and biomass accumulation for the ecosystem are known with some precision, a budget of these components may be used to estimate unknown parameters such as net weathering release or cationic denudation (Likens et al., 1977). This is directly analogous to what we did with the two atmophilic elements N and S to calculate net meteorologic inputs of gases and aerosols. In that case it was possible to use the budget method because the concentrations of these two elements in the weathering substrate were negligible or very small and the other components were known. Using the budget method (Figure 1-14), we can estimate net weathering release for each element by the relationship (meteorologic input + biomass immobilization − streamwater output = weathering). Weathering for Ca calculated by the budget method is a reasonable estimate of the total amounts actually released by weathering, since calcium-rich minerals are thought to be the least stable and calcium is not thought to be significantly involved in the formation of secondary minerals at Hubbard Brook (Johnson et al., 1968; Likens et al., 1977). Assuming that calcium is completely extracted from primary minerals in the weathering process, we calculate that about 1500 kg/ha of rock or till are weathered each year (Likens et al., 1977).

Net weathering release calculated by the budget method does not distinguish between the processes of differential rates of release and formation of secondary minerals. As we have already discussed, aluminum, silicon, and iron give strong indications of being involved in secondary mineral formation. Hence, the actual weathering rates for these elements are probably higher than those estimated by difference. In acid soils, phosphorus is thought to form a variety of insoluble compounds through reactions with soluble and hydrous oxides of Fe, Al, and Mg (Brady, 1974); hence, the actual weathering rate of phosphorus is likely to be higher than estimates of net release of phosphorus based on the budget method. Potassium and Mg may be involved to a lesser extent in secondary mineral formation and, along with Na, may have a differential rate of release from the weathering substrate at Hubbard Brook (Johnson et al., 1968; Likens et al., 1977).

### Relative Importance of Weathering and Meteorologic Input

The relative importance of the various sources for biologically important nutrients for the aggrading ecosystem are given in Table 2-8. Weathering is the major source of several elements—Fe, Ca, K, Mg, Na, Al, and P. Meteorologic input is the most important source of N and S and a secondary source of Na, Mg, K, and Ca.

## CIRCULATION AND RETENTION OF NUTRIENTS

A fundamentally important characteristic of the aggrading forested ecosystem is its ability to store and recycle nutrients. This frugal husbandry of vital resources assures the constant growth and development of the ecosystem during this period. Losses are minimal, and available and stored resources accumulate. This may be illustrated most clearly by an examination of the nitrogen cycle.

### The Nitrogen Cycle

The biogeochemistry of nitrogen requires special consideration, since this element plays a key role in the metabolism of the ecosystem. The ecosystem accumulates nitrogen throughout the Aggradation Phase >15 through 170 years after clear-cutting (Figure 2-10). During the first third of the period, the rate of net accumulation is high, ca. 20 kg/ha-yr, but accumulation tapers off to about 5.0 kg/ha-yr during the last 30 years. The change in rate is due to the coincidence of growth in the living biomass and forest floor during Years 15 to 75 and the fact that accumulation in living biomass alone sustains the increase in the standing crop after Year 75 and rate of living biomass accumulation begins to level off after Year 140. It is interesting to note that dead wood has little direct influence on the ecosystem's standing crop of nitrogen.

As we have already discussed, there are two possible sources of nitrogen to sustain accumulation—the input of nitrogen in bulk precipitation and the fixation of gaseous nitrogen by microorganisms within the ecosystem. We can speculate about the relative importance of these sources throughout the period by using detailed biogeochemical data from small watersheds at Hubbard Brook. The input of nitrogen in bulk precipitation has averaged 6.5 kg/ha-yr. This provides the upper limit of nitrogen that might be accreted from precipitation; however, some nitrogen is inevitably lost in drainage water, and if we subtract the 11-yr

**Table 2-8.** Relative Importance of Various Sources of Nutrients for the Hubbard Brook Experimental Forest[a]

| | Nutrient Source (%) | | |
|---|---|---|---|
| Nutrient | Precipitation Input | Net Gas or Aerosol Input | Weathering Release |
| Iron | —[b] | —[b] | 100 |
| Calcium | 9 | — | 91 |
| Potassium | 11 | — | 89 |
| Magnesium | 15 | — | 85 |
| Sodium | 22 | — | 78 |
| Nitrogen | 31 | 69[c] | —[b] |
| Sulfur | 65 | 31[d] | 4 |

[a]*Data from Likens et al., 1977.*
[b]*Nearly zero.*
[c]*Nitrogen fixation.*
[d]*Impaction and gaseous absorption.*

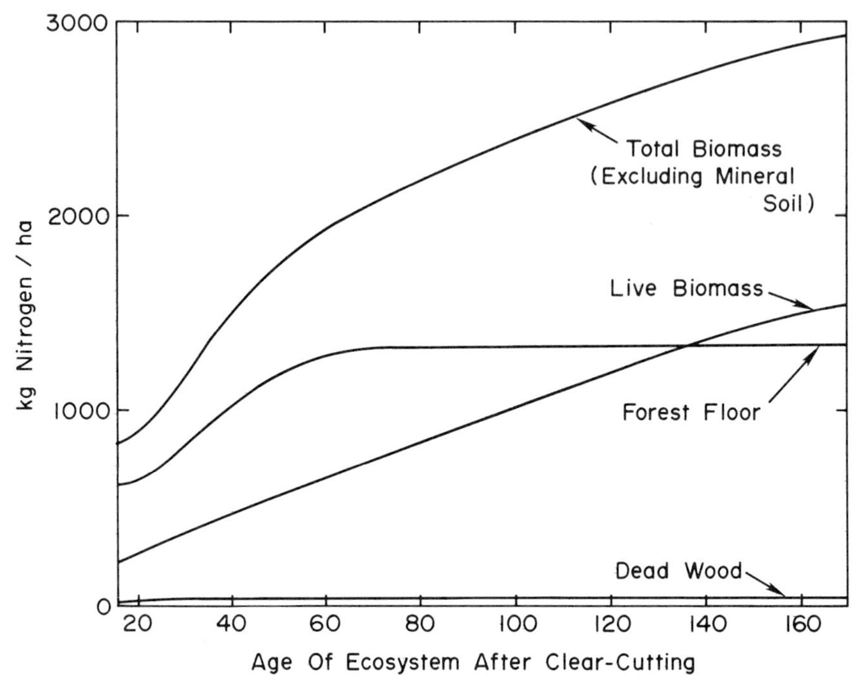


**Figure 2-10.** Estimated accumulation of nitrogen during the Aggradation Phase based on living and dead biomass projections multiplied by appropriate nitrogen concentrations. The estimate of total nitrogen does not include nitrogen in the mineral soil.

average stream-water loss, 4.0 kg/ha-yr, the net precipitation contribution to nitrogen accumulation is about 2.5 kg/ha-yr (Table 2-6). If we subtract this net contribution from the yearly net accumulation of nitrogen (Figure 2-10), we obtain an estimate of the net accumulation due to nitrogen fixation. This number varies from about 28 kg/ha-yr in Years 20 to 40 to about 2 kg/ha-yr during Years 140 to 170.

However, there is a problem with this method of calculation since it cannot take into account denitrification. Thus, it is conceivable that denitrification fluctuates greatly during the aggrading period, whereas nitrogen fixation does not; this would cause the balance within the system to be affected more by changes in denitrification than by changes in nitrogen fixation. Because dead wood is an important site of nitrogen fixation, it is interesting to note that the standing crop of dead wood (Figure 2-1) remains about the same from Year 50 to 170. Difficulties involved in long-term evaluations of nutrient behavior will be discussed more fully in Chapters 5 and 6.

Several important features of the nitrogen cycle emerge from the overall pattern as seen in a 55-yr-old ecosystem at Hubbard Brook (Figure 2-11). These are: (1) 70% of the nitrogen added to the ecosystem is added by nitrogen fixation, 30% is added in precipitation, and little, if any, is added by weathering; (2) of the estimated 20.7 kg/ha entering the system, about 80% is held or accreted within the ecosystem; (3) of the estimated 83.6 kg/ha of nitrogen compounds added to the inorganic nitrogen pool, only 5% leaks out of the ecosystem; and (4) of the 119 kg of nitrogen estimated to be used in growth processes of the plants, 33% is withdrawn from storage locations within plant tissues in the spring and utilized in growth and a like amount is withdrawn from the leaves and stored in more permanent tissues shortly before leaf senescence in the fall (Ryan, 1978).

The latter point is of particular significance since it suggests that a good portion of the early spring growth is sustained by nitrogen (and phosphorus) withdrawn from the nutrient pool within living plant tissue. This internal reserve gives the plants some independence from soil sources during the critical early-growth period (Ryan, 1978). The internal resorption pool would be almost completely destroyed in clear-cutting!

Nitrogen incorporated in organic compounds within the aggrading ecosystem is ultimately decomposed in a number of steps to ammonium. Ammonium ions may be strongly held on cation-exchange sites, fixed on organic matter or clay, or taken up directly by plants. Ammonium, along with some other nitrogenous compounds, may be used as a substrate for the process of nitrification—the oxidation of nitrogen by a variety of chemoautotrophic and heterotrophic microorganisms with the production of nitrate and hydrogen ions (Table 2-9). Nitrate is very soluble and may be easily leached from the ecosystem if it is not taken up by plants.

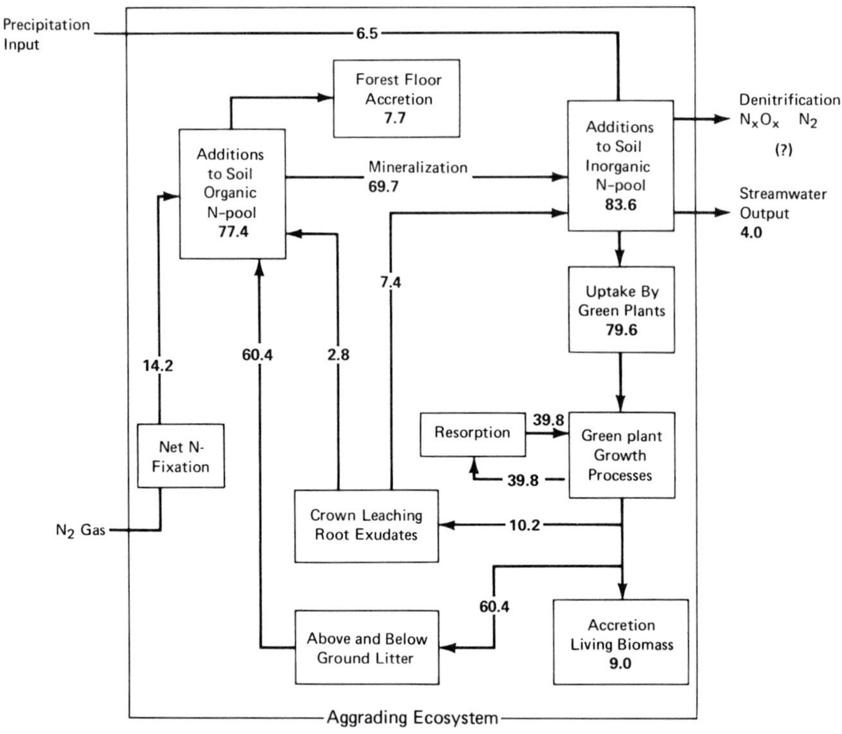


**Figure 2-11.** Major nitrogen transfers into, out of, and within a 55-yr-old aggrading northern hardwood ecosystem. Numbers are kilograms of nitrogen per hectare per year (Bormann et al., 1977).

Ammonium, on the other hand, as a cation, is more tightly retained in the soil profile.

A major question regarding nitrogen flow in the aggrading ecosystem is the degree to which ammonium is nitrified to nitrate. The forest floor of some podzol soils, particularly under conifers, may show little tendency to nitrify (Romell and Heiberg, 1931; Corke, 1958), and it may be presumed that ammonium is the major nitrogen ion utilized by the vegetation.

**Table 2-9.** Chemoautotrophic Nitrification[a]

| | |
|---|---|
| Step 1 | |
| $NH_4^+ + 1\frac{1}{2}\ O_2 \rightarrow NO_2^- + 2H^+ + H_2O$ | 66 Kcal |
| Step 2 | |
| $NaNO_2 + 1\frac{1}{2}\ O_2 \rightarrow NaNO_3$ | 18 Kcal |

[a] *After Alexander (1961).*

Many other podzol and podzolic soils, particularly under northern hardwood forests, have forest floors that have a strong tendency to nitrify when a sample is removed and kept moist in the laboratory (Romell and Heiberg, 1931). In these latter soils, however, there is little information on the amount of nitrification that occurs within the intact forest ecosystem.

Soils under aggrading northern hardwoods at Hubbard Brook fit into the latter category since samples brought into the laboratory rapidly nitrify (J. Duggin, personal communication); Smith et al. (1968) have isolated chemoautotrophic-nitrifying bacteria from these soils. Application of a chemical inhibitor of autotrophic nitrification reduced the rate of nitrification in these laboratory samples by about 50%. This indicates that nitrification is not simply the result of chemoautotrophic nitrification but probably has a heterotrophic component as well (J. Duggin, personal communication).

Levels of exchangeable ammonium and nitrate measured throughout the year at Hubbard Brook in the soil and in soil placed in plastic bags and replaced in the soil (the Solling Method) suggested that no more than 10 to 20% of the ammonium is nitrified to nitrate during the year (Melillo, 1977). Dominski (1971) estimated that about 16% of the ammonium was nitrified and that about 12-kg/ha of $NO_3$-N were produced each year. Thus, despite the strong tendency of samples from the forest floor at Hubbard Brook to undergo nitrification in the laboratory, it is apparent that under field conditions only a small proportion of the nitrogen cycling within the forest is converted to nitrate.

From this, we conclude that, although a strong potential for nitrification exists, that potential is not realized in the intact aggrading ecosystem. Some factor such as competition for ammonium ions between green plants and nitrifying microorganisms, or some set of factors, partially inhibits the process. Another possibility is allelopathic inhibition of nitrifying organisms by the dominant vegetation of the ecosystem (Rice, 1974). Studies by Melillo (1977) indicate that root exudates may cause some allelopathic inhibition of nitrification in the soil at Hubbard Brook.

The partial inhibition of nitrification in the intact aggrading ecosystem has interesting energetic implications. Nitrification is an oxidative exoergic process that liberates energy originally built into the nitrogen atom when it was reduced. Since it is in a more oxidized state, the utilization of a nitrate ion by a green plant requires a considerably greater expenditure of the plant's energy (to reduce the nitrogen) than utilization of an ammonium ion, which is already in a reduced state. This suggests that inhibiting the nitrification process might result in an increase in net primary productivity. We can make a rough calculation of this as follows: In Figure 2-11, we estimate that approximately 70-kg of nitrogen or organic compounds are mineralized to ammonium each year. If all of this were nitrified and taken up by green plants as nitrate, the plants would have to expend at least $4.2 \times 10^5$ Kcal/ha to reduce the nitrogen to a biochemically usable form. This amounts to about 1% of the net primary production of the 55-yr-old aggrading ecosystem (Table 2-1).

Nitrogen relationships within the aggrading ecosystem are reflected in a highly reproducible pattern in the stream water draining these systems. Nitrate concentrations, which are always low, i.e., <6 mg/liter, fall to their lowest levels in late May when vegetation growth begins (Figure 2-3) and remain low until October when the vegetation becomes dormant. Thereafter, concentrations gradually rise to a maximum in late winter and early spring. This pattern is related to the heavy utilization of nitrogen by the biological fraction of the ecosystem during its active season (Johnson et al., 1969). The stream-water pattern provides a useful monitor of the nitrogen cycle within the ecosystem, and, as we shall see, this pattern is wholly disrupted by clear-cutting.

Thus, we conclude that within the aggrading ecosystem most nitrogen compounds are decomposed to ammonium, which is directly taken up by green plants. The process of nitrification, although it occurs, does not reach its full potential because of the action of unknown factors or inhibiting substances within the aggrading ecosystem. Stream water draining the aggrading ecosystem reflects these conditions and shows relatively small losses of nitrate and ammonium from the ecosystem. On this basis, we may consider the cycling of nitrogen within the aggrading system to be tight. Specifically, only about 5% of the nitrogen added to the inorganic pool (Figure 2-11) is lost in drainage water.

## Summary

1. Our studies of the aggrading forested ecosystem show that the flow of water, nutrients, and energy is highly regulated by biotic and abiotic components of the ecosystem.
2. Stability of the aggrading ecosystem is characterized by:
   a. modest but rather constant excess of primary production over decomposition, which produces a predictable rate of accumulation of living and dead biomass;
   b. close regulation of both the chemistry of drainage waters and erodibility; and
   c. an ability to exert considerable control over intrasystem aspects of the hydrologic cycle, such as losses of water by evapotranspiration.

   These processes, integrated within limits set by climate, geology, topography, biota, and level of ecosystem development, determine the size of nutrient reservoirs and produce nutrient cycles typified by minimum outputs of dissolved substances and particulate matter and by maximum resistance to erosion.

3. The ability of the biotic fraction of the ecosystem to utilize external energy sources to convert liquid water to vapor through transpiration and interception results in the conservation of nutrients by its effect on the volume of flow and on discharge rates.
   a. The reduction of streamflow has the direct effect of holding nutrients within the ecosystem because, in the aggrading ecosystem, concentrations of dissolved substances in drainage water are highly predictable and are only modestly affected by flow rate. Hence, the output of dissolved substances is closely related to the amount of streamflow exiting from the ecosystem. Factors regulating the concentrations of nutrients in drainage water are only partially understood, but the following may be important: quantity and quality of inputs, rates of weathering, micrometeorological conditions at the level of the forest floor which regulate soil chemistry, and allelopathic control of soil microorganisms by living vegetation.
   b. Particulate matter losses are a nonlinear direct function of discharge rates. In the aggrading ecosystem, transpiration and interception can damp discharge rates by lowering the quantity of water stored within the system. Thus, a heavy rain when the storage volume of the system is relatively empty will produce relatively little flow and modest discharge rates so long as storage capacity is unsatisfied. Transpiration tends to keep hydrologic storage at a minimum during the growing season, and streamflow during summer months is typically low even though precipitation is evenly distributed throughout the year.
   c. Particulate matter losses are a function not only of discharge rate but also of erodibility or the capacity of the ecosystem to resist erosion. Although geologic and other abiotic factors are important in determining the erodibility of an ecosystem, biota and its inorganic debris exerts considerable control over this phenomenon.
4. The effectiveness of all of these mechanisms in maintaining the integrity of the aggrading ecosystem against external forces of geological degradation is seen in the fact that the 55-yr-old forested ecosystems at Hubbard Brook have average gross dissolved-substance and particulate-matter losses of 13 and 2.5 t/km$^2$-yr, respectively. These losses constitute only a small fraction of the elements stored in the various compartments of the ecosystem. This condition exists despite the location of the Hubbard Brook ecosystem-watersheds on relatively steep slopes subject to an average rainfall of 130 cm/yr.
5. As an example of nutrient cycling in the aggrading ecosystem, the nitrogen cycle is discussed. Salient features are:
   a. The system is accumulating nitrogen both from precipitation input and nitrogen fixation.

b. The rate of accumulation changes with time and is not simply related to biomass accumulation.
c. Nitrogen cycling is tight—only about 5% of the mineralized nitrogen is lost as geologic output.
d. The soils under the northern hardwood ecosystem have a strong potential to nitrify, but under the aggrading condition relatively little nitrification occurs, and most nitrogen is cycled as ammonium.

## CHAPTER 3

# Reorganization: Loss of Biotic Regulation

The Reorganization Phase in the northern hardwood developmental sequence is characterized by drastic changes in hydrologic, energetic, ecological, and biogeochemical processes that in the Aggradation Phase were fairly constant and predictable. Rates of net primary production, transpiration, and nutrient uptake registered by plant growth during the first growing season after cutting are far below levels in the uncut forest. There are also rapid and marked increases in internal ecosystem parameters like decomposition, nitrification, available soil moisture, and soil temperature and export parameters like summertime streamflow, nutrient concentration in stream water, and erosion. Cutting also imposes immediate and significant shifts in stores of nutrients and organic matter in the living (loss) and dead (gain) biomass compartments of the ecosystem.

The Reorganization Phase begins immediately after clear-cutting and concludes when rates of internal ecosystem processes and ecosystem exports such as water, particulate matter, and nutrients approach levels characteristic of the aggrading forest ecosystem (Figure 1-10). Not all processes return to former levels simultaneously, so the termination of the Reorganization Phase is chosen as the point at which both living and dead biomass are accumulating within the ecosystem. This occurs about 15 years after clear-cutting when the rate of decline of organic matter in the forest floor (Figure 1-10) reaches a null point and the forest floor begins to accumulate biomass.

During Reorganization, there is a loss of biotic regulation. Hydrologic and biogeochemical parameters, which were highly regulated in the aggrading ecosystem, are grossly altered and for a short time appear to be "out of control." However, this apparent loss of control may be considered part of a complex feedback mechanism that acts to return the ecosystem to predisturbance levels of productivity and biogeochemistry. The loss in biotic regulation results in a temporary increase in resource availability, i.e., available water and nutrients, as well as increased radiant energy at the forest floor level. The change in resource availability is coupled with the activation of reproductive and growth strategies that allow plant species to exploit the new conditions. Rapid growth results in a return of biotic regulation over ecosystem processes to precutting levels.

We attempt to deal with these kaleidoscopic changes in the next three chapters. In the remainder of this chapter, we shall consider ecosystem responses *in the absence of vegetative regrowth.* This procedure allows us to see more clearly the behavior of the heterotrophic components of the ecosystem immediately after the destruction of vegetation that begins the Reorganization Phase. Chapter 4 will be concerned with species strategies and community dynamics during both the Reorganization and Aggradation Phases, and Chapter 5 will be concerned with the recovery of biotic regulation that occurs as the various species' strategies respond to new conditions of resource availability created by clear-cutting.

## A DEFORESTATION EXPERIMENT

More than a decade ago we began an experiment within the Hubbard Brook Experimental Forest to evaluate the responses and recovery of a forested ecosystem to a disturbance such as clear-cutting. In cooperation with the U.S. Forest Service, all of the trees on a small watershed of 15.6-ha (W2) were cut during the autumn of 1965 (Figure 1-11). No wood products were harvested or removed, and no roads or skid trails were built. All trees were felled in place, and the slash was leveled to within 1.5-m of the ground (Figure 3-1). Thus disturbance of the forest floor was minimized, and post-cutting erosion rates could be used to assess the response to deforestation rather than the effect of mechanical damage to the soil.

To separate the effects of nutrient uptake by regrowing vegetation from the effects on nutrient release resulting from deforestation, herbicide (Bromacil and 2,4,5-T) was applied during three growing seasons to keep the watershed bare and suppress vegetative regrowth. This tended to maximize the loss of nutrients from the ecosystem, which was useful since we use nutrient output as an indirect measure of activities with the system. At the inception of the experiment, we had no guide as to expected results, and minor changes in nutrient output might have been difficult to measure. Analogous to experiments in cell biology in which genetic blocks are placed on biochemical pathways and accumulated products are studied for clues to cell biochemistry, our experiment

**Figure 3-1.** Watershed 2 during the spring of 1966. All trees greater than 2 cm were cut and dropped in place during November and December of 1965.

(deforestation) placed blocks on two major ecosystem pathways, i.e., uptake of water and nutrients and litter production by green plants. Hydrologic and chemical output was compared with an aggrading ecosystem adjacent to W2 to discern differences in export behavior, to look for clues to internal biogeochemical function, and to study recovery mechanisms (Likens et al., 1970; Pierce et al., 1970; Bormann et al., 1974). Prior to cutting, both watersheds supported second-growth northern hardwood forests of the same age and history and had been monitored for hydrology (9 yr), losses of dissolved substances (3 yr), and particulate matter (1 yr).

After 3 years of deforestation, W2 was allowed to revegetate, beginning in the summer of 1969, and continues to do so. Revegetation studies by W. A. Reiners of Dartmouth College and continuing biogeochemical studies during this period allow a detailed study of the processes of ecosystem recovery after experimentally prolonged deforestation.

## Energy Flow

Deforestation and suppression of regrowth greatly modified the ecosystem's capacity to regulate the flow of radiant energy. Major effects resulted from the virtual elimination of a summertime reflective surface (canopy) about 20-m above the soil and the elimination of transpiration. Pierce et al. (1970) report that streamflow increased an average of 28.6-cm/yr after cutting and that the great bulk of increase was due to the elimination of transpiration. Approximately 585 calories are required to transpire 1-ml of water; hence, $1.67 \times 10^9$ Kcal/ha-yr of energy, previously used to transpire water (Table 2-1), reached the vicinity of the forest floor in the form of radiant energy. This, coupled with energy previously reflected, resulted in a substantial heat load on the forest floor. These changes in energy flow had the dual effect of increasing summertime soil temperature while at the same time increasing soil moisture. The latter effect resulted from the elimination of transpiration.

## Hydrology

The elimination of transpiration had a pronounced effect on ecosystem hydrology. Hydrologic output generally was influenced by variations in precipitation, including particularly subnormal rainfall during the summer of 1968; nevertheless, a strong seasonal pattern emerged (Hornbeck et al., 1970). Summertime streamflow was increased dramatically by deforestation (Table 3-1), whereas winter runoff was affected very little except for timing.

Monthly precipitation during the growing season averages 115 mm (Likens et al., 1977). Potential evapotranspiration by the forest usually exceeds this amount, and soil water deficits progressively develop during the growing season. Precipitation rarely is sufficient to offset these deficits, and high streamflows are seldom recorded during the growing season in the aggrading forest (Figure 3-2). In the deforested watershed, soils remain relatively wet because of reduced evapotranspiration, and little storage capacity is available. Differences in storage of available soil water between forested and deforested watersheds can amount to 50–75-mm. Consequently, summer rain on the already wet soils of the deforested watershed quickly becomes streamflow and, in some instances, generates large streamflows. During one large midsummer storm (29–30

**Table 3-1.** Effect of Deforestation on Seasonal Streamflow

| | Streamflow Increase (mm)[a] | | |
|---|---|---|---|
| Season | 1966–67 | 1967–68 | 1968–69 |
| 1 June–31 October | 332 | 248 | 164 |
| 1 November–31 May | 4 | 15 | 72 |
| Total[b] | 336 | 263 | 236 |

[a] *Increase as compared to values predicted for the vegetated condition based on regression equations developed during the 8-yr pretreatment period (from Pierce et al., 1970).*
[b] *Totals for any year vary slightly from those given in the text which were computed with separate regressions based on total flow for the year.*

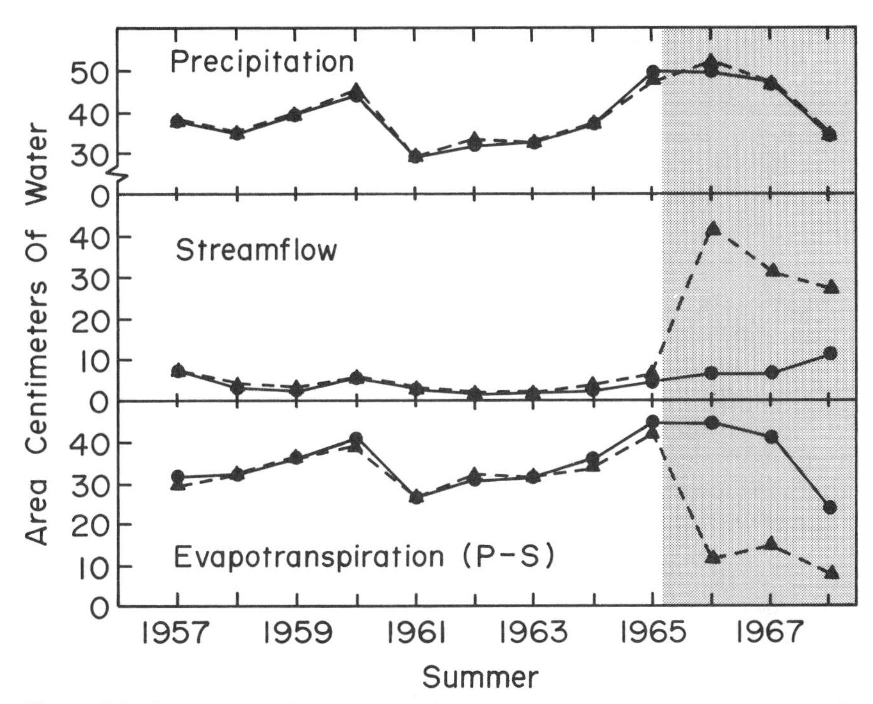


**Figure 3-2.** Summer hydrology, 1 June through 30 September. (●—●) Watershed 6, aggrading forest ecosystem; (▲---▲) Watershed 2, deforested during November and December of 1965 and maintained bare by herbicide treatments. Shaded area represents period when Watershed 2 was devegetated.

July 1969), precipitation of 137 mm generated an instantaneous peak flow of 305-mm/day on the deforested watershed as compared to 100-mm/day for the forested watershed (Hornbeck, 1973b).

During the dormant season, soil water storage is about the same on both watersheds; hence there is generally little difference in amount of streamflow or in instantaneous streamflow peaks, except as related to difference in snowmelt.

Predictions of annual streamflow for W2, had it not been cut, were made from streamflow regressions of W2 on an adjacent forested watershed. These regressions were established prior to the cutting of W2. Annual streamflow during the 3 years of devegetation increased by 40, 28, and 26% (346, 273, and 240-mm, respectively) over the values predicted if W2 had not been cut (Pierce et al., 1970). These increased yields of water are very important to the biogeochemical relationships, since stream water is the major vehicle for nutrient loss of dissolved substances and particulate matter.

Hydrologic effects of devegetation are not limited to quantities of water. Patterns of discharge and storm peaks also are affected. The discharge of snowmelt during the spring was advanced by several days. Peak discharges also were more common and more pronounced during the summer in the deforested watershed owing to the wetness of the soil and the rapidity with which precipitation became streamflow. These responses are most important since storm peaks are exponentially related to erosion and transport of particulate matter (Bormann et al., 1974).

However, to emphasize the ever-mind-boggling complexity of natural ecosystems, this pattern was reversed during the winter of 1971–72 when the thaw period included rain. The stream channel draining the deforested watershed was deeply frozen owing to its greater exposure to the sky, while streams draining forested watersheds were only lightly frozen. During this particular winter thaw period, the stream in the deforested watershed remained partially frozen, while the streams in the forested watersheds completely thawed. The ensuing flood on thawed streambeds produced massive outputs of particulate matter from forested watersheds, but only modest output was recorded from the devegetated watershed (Bormann et al., 1974).

## Dissolved Substance Concentrations in Stream Water

The highly predictable relationship between streamflow and concentrations of dissolved substances that is characteristic of the aggrading forest ecosystem (Johnson et al., 1969) was rapidly and markedly altered by devegetation (Likens et al., 1970). With the onset of the first growing season after cutting, stream-water concentrations of almost all ions rose considerably above those in the forested ecosystem (Figure 3-3). This occurred relatively quickly, i.e., within several weeks, and was coincident with the flow of much greater quantities of water through the deforested ecosystem. For the 3 water-years during which revegetation was suppressed by herbicides (1 June 1966 to 31 May 1969), average stream-water concentrations of ions from the devegetated system exceeded those of the forested ecosystem by the following factors: $NO_3^-$, ×40; $K^+$, ×11; $Ca^{2+}$, ×5.2; $Al^{3+}$, ×5.2; $H^+$, ×2.5; $Mg^{2+}$, ×3.9; $Na^+$, ×1.7; $Cl^-$, ×1.4; and dissolved silica, ×1.4. Ammonium concentrations increased, but in absolute amounts the change between the two watersheds was small. Bicarbonate concentrations decreased 7-fold, and sulfate concentrations declined by a factor of 1.5 owing to complicated interactions involving the following factors: dilution from greater amounts of stream water leaving the ecosystem, an inverse relationship to $NO_3^-$ concentrations, a potential increase in anaerobic sulfate reduction (Likens et al., 1970), and a possible reduction in atmospheric input from absorption or impaction owing to canopy removal.

For most ions, stream-water concentrations reached a peak during the second year after cutting and declined during the third year (Figure 3-3), even though the ecosystem remained bare.

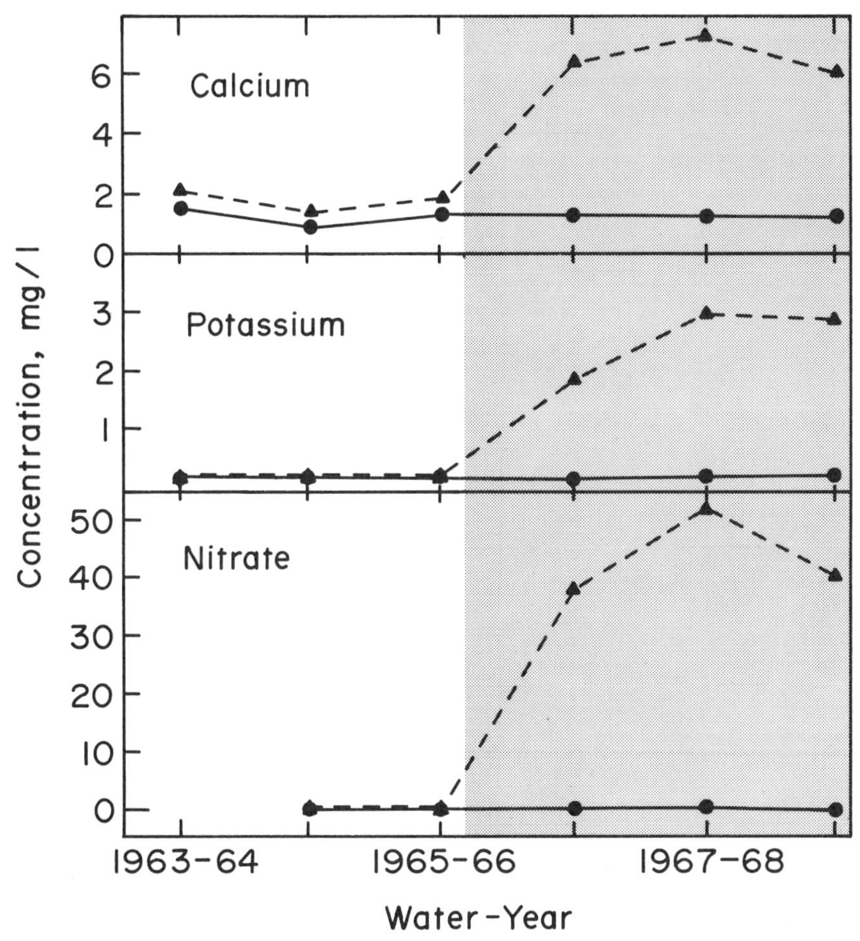


**Figure 3-3.** Annual weighted average of dissolved ions in stream water: (●——●) Watershed 6; (▲---▲) Watershed 2. Shaded area represents period when Watershed 2 was devegetated.

## Net Losses of Dissolved Substances

The pattern of tight nutrient cycling and the highly predictable relationship between the annual output of dissolved substances and annual streamflow (Figure 2-9) characteristic of the aggrading forest ecosystem was markedly disrupted by deforestation. During the 3-yr period of deforestation, total net losses of dissolved substances were approximately eight times greater than those from the forested ecosystem (Table 3-2). Increases in loss were as follows: $K^+$, ×20; $Ca^{2+}$, ×8.6; $Al^{3+}$, ×7.0; $Mg^{2+}$, ×6.0; and $Na^+$, ×2.5. Based on a comparison of gross losses, N loss was increased by a factor of approximately 160. We conclude from this that the ecosystem responds to enforced devegetation by dumping substantial quantities of nutrients into drainage water.

**Table 3-2.** Annual Net Loss or Gain of Dissolved Substances for an Aggrading Forest (W6) and a Deforested Ecosystem (W2), from June 1966 to June 1969[a]

| | Net Loss or Gain (kg/ha)[b] | |
|---|---|---|
| Element | W6 | W2 |
| Ca | −9.0 | −77.7 |
| Mg | −2.6 | −15.6 |
| K | −1.5 | −30.3 |
| Na | −6.1 | −15.4 |
| Al | −3.0 | −21.1 |
| $NH_4$-N | +2.2 | +1.6 |
| $NO_3$-N | +2.3 | −114.1 |
| $SO_4$-S | −4.1 | −2.8 |
| Cl | +1.2 | −1.7 |
| $HCO_3$-C | −0.4 | −0.1 |
| $SiO_2$-Si | −15.9 | −30.6 |
| Total | −36.9 | −307.8 |

[a] *Stream-water output minus meteorological input equals net change. W2 was deforested in November–December 1965.*
[b] *Weighted average.*

Nutrient losses reached a peak in the second year even though devegetation was continued for a third year. This phenomenon is more closely governed by nutrient concentrations in the stream water or soil solution than by amounts of water flowing through the soil. The range in total annual streamflow from W2 during the 3 years was only 6% (Pierce et al., 1970), while the range in concentration of total dissolved substances was 28% (61.6-mg/liter in 1966–67, 78.7-mg/liter in 1967–68, and 63.5-mg/liter in 1968–69).

These data suggest a very important point. The decline in concentration during the third year of the deforestation experiment (1968–69) apparently resulted from the progressive exhaustion of the supply of easily decomposable substrate present in the ecosystem at the beginning of devegetation. In other words, the amount and type of decomposable substances apparently limited nutrient outputs after only 2 years.

## Factors Governing Dissolved Substances

Devegetation resulted in a rapid rise in the concentration of dissolved substances in the soil-solution–stream-water continuum and the export of chemicals from the ecosystem. It is apparent that the elimination of nutrient uptake by plants at the outset of the devegetation experiment played a major role in changing soil and stream-water chemistry. Available nutrients generated by decomposition and mineralization apparently accumulated in the soil and were flushed from the ecosystem. Recently, Vitousek (1977) suggested that the increased loss of nitrate that occurred in our devegetation experiment and that follows clear-cutting could be *wholly explained* by the elimination of uptake by plants. We think this is a much too simple explanation. Devegetation not only eliminates nutrient uptake but results in a whole array of changes which affect soil chemistry and stream-water losses from the ecosystem.

We now consider the role that these other factors play in altering the basic relationships which control the output of dissolved substances after devegetation.

*Origin of Nutrients Flushed from the Ecosystem.* Likens et al. (1970) suggested that changes in weathering rates had a minimal impact on the biogeochemical behavior of the ecosystem after cutting. Prior to cutting, the Ca/Na ratio in the stream water of W2 was 1.6/1.0. For the 2-yr period after cutting, the ratio climbed to 4.8/1.0. More significantly,

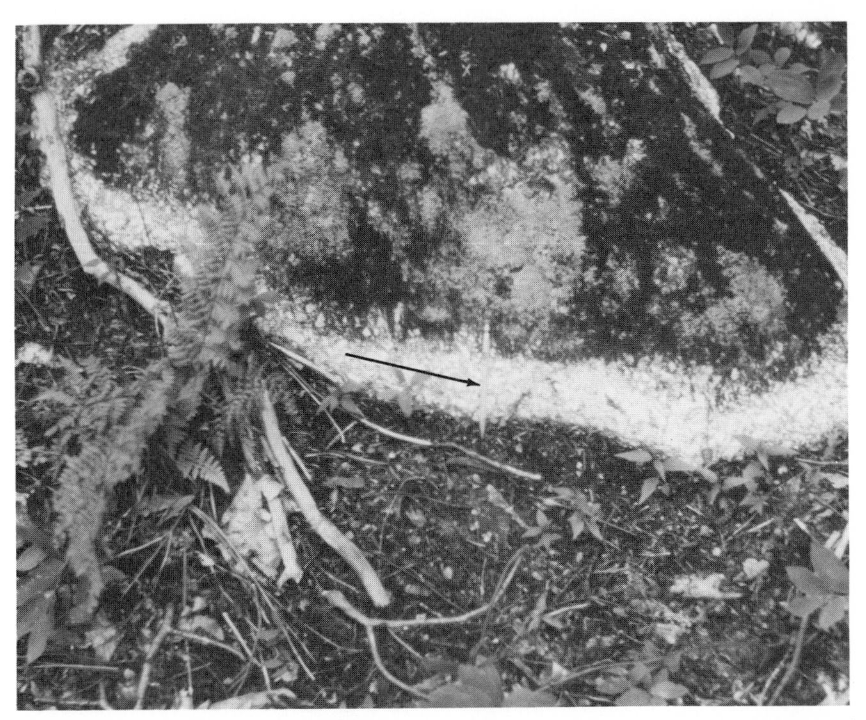

**Figure 3-4.** A large rock on Watershed 2 shown during the second summer of devegetation. The whitish area represents weathered rock exposed as the forest floor declined in depth owing to both decomposition and compaction. During the 3-yr devegetation period the forest floor lost about 23% by weight and about 3-cm in depth (Dominski, 1971). Note pencil in center of photo for scale.

the Ca/Na ratio for "excess" ions, or the amount added above the undisturbed condition, was 7.4/1.0. Thus, the net effect of devegetation was to produce soluble calcium differentially within the ecosystem, mostly from the forest floor (Figure 3-4). The forest floor is rich in calcium relative to sodium (53/1; Gosz et al., 1976), in contrast to average bedrock for Hubbard Brook (0.9/1.0 Ca/Na; Johnson et al., 1968), which suggests that most of the "excess" ions produced are derived from decomposition of organic materials rather than accelerated weathering. In fact, the presence of high concentrations of ions in the soil solution, which are also end-products of weathering, probably reduces weathering of some minerals.

*Decomposition and Mineralization.* In the absence of nutrient uptake by green plants, a sizable proportion of the end products of decomposition and mineralization were flushed out of the ecosystem and thus provided an indirect measure of these processes. For example, a net amount of 337-kg/ha of dissolved inorganic nitrogen was flushed out of W2 during the 3-yr period, which suggests that at least that much nitrogen was released by decomposition and mineralization.

Dominski (1971) sampled the soil in the devegetated watershed at yearly intervals and found that organic matter losses after 3 years were 10,800 kg/ha from the forest floor (~23% of the total; Table 2-3; Figure 3-4) and 18,900 kg/ha from the soil body as a whole. Not only were nutrients mineralized from these organic compounds, but nutrients held on organic exchange surfaces were released when the surface was decomposed. At the end of 3 years, cation-exchange capacity in the 0 to 15-cm layer had decreased from 172 to 137 kEq/ha, a reduction of about 20%. For the soil body as a whole, Dominski estimated a 3-yr net loss of nitrogen of 472-kg/ha. All things considered, this figure compares remarkably well with the 337-kg/ha of dissolved inorganic nitrogen measured in stream-water export.

We propose that the deforestation treatment at first markedly increased the rate of decomposition over that expected in an uncut forest. Dominski (1971) reported soil temperature differences as great as 8.5 and 3.0°C in the 0 to 2.5-cm and 10 to 15-cm soil layers, respectively, between the devegetated and forested ecosystems. Coupled with the higher temperatures during the growing season is a greater abundance of soil moisture resulting from the elimination of transpiration, as mentioned previously. These factors, together with the richer concentrations of dissolved nutrients in the soil solution, suggest the potential for marked elevation in the rate of decomposition.

However, comparative changes in soil decomposition rates between devegetated and forested ecosystems are complicated to measure. The decomposition substrate is composed of a vast variety of organic materials, ranging from easily decomposable substances to substances highly resistant to decomposition. At the beginning of the first growing season after cutting, decomposition substrates in the cut and forested ecosystems were quite similar. Gradually, however, marked differences developed as the decomposition substrate was continually renewed by litterfall in the forested ecosystem, while the substrate in the devegetated system progressively became enriched in substances more resistant to decomposition. Hence, a comparison of decomposition rates, as measured by carbon loss, had less meaning during subsequent years as the substrate in the devegetated system became relatively more resistant to decomposition. The decomposition rate derived from Dominski's loss of organic matter during the first year for the soil body as a whole in the devegetated watershed (12,000-kg/ha) is approximately double that estimated for the aggrading forest (6500-kg/ha, corrected for growth of the forest floor; Gosz et al., 1976). Organic matter losses for the last 2 years (Dominski, 1971) are well below the estimate for decomposition in the undisturbed forest (Gosz et al., 1976). However, if we use nitrogen losses as an indirect measure of decomposition, a different picture emerges. Using either annual stream-water losses of inorganic nitrogen or Dominski's (1971) estimates of annual losses of nitrogen from the soil profile in W2, maximum losses occur in the second year. For example, stream-water losses during the first year after cutting are about 1.4 times greater than the 70-kg of nitrogen estimated to be mineralized annually in the aggrading forest (Figure 2-11; Bormann et al., 1977); second-year losses are 2.0 times greater; and third-year losses are 1.4 times greater.

These data indicate that attention must be given both to decomposition, as measured by loss of carbon from the ecosystem, and to the relationship of decomposition to the loss of other nutrients such as nitrogen and phosphorus. Gosz et al. (1973) has pointed out that in the cycle of decomposition at Hubbard Brook there are critical carbon/element and element/phosphorus ratios for many nutrients. These ratios reflect nutrient demand by decomposers, and only after ratios are reduced below the critical levels can all of the end products of mineralization appear in drainage water. Data on losses of nutrients in stream water (Likens et al., 1970) and from the soil (Dominski, 1971) show that losses reached a peak during the second year of devegetation and declined thereafter. It seems probable that supplies of easily decomposable nutrients became limiting to this process after the second year.

*Effects of Devegetation on Nitrification.* Destruction of vegetation rapidly accelerates the process of nitrification. Within a few weeks of the onset of the first growing season after cutting, nitrate concentrations in stream water draining the deforested ecosystem rose far above any previously reported for an aggrading forest (Likens et al., 1970). If we use 13-kg of N/ha as the estimate of the total annual nitrification within the aggrading forest (Dominski, 1971; Melillo, 1977) and the annual output of $NO_3^-$ in stream water as the estimate of nitification in the deforested system, we can estimate that the annual rate of nitrification was increased by about 7, 11, and 9 times during the 3 years after deforestation.

The process of nitrification not only produces the highly soluble anion $NO_3^-$ but also produces two hydrogen ions for each nitrate anion. The hydrogen ions can exchange for other cations held on the exchange surfaces and bring about the solubilization of cations previously bound to those exchange surfaces. Likens et al. (1970) reported that the changes in stream-water chemistry after devegetation could no longer be fully explained by the dilution–concentration model that worked so well in the aggrading forest (Johnson et al., 1969). Rather, concentrations of $Mg^{2+}$, $Ca^{2+}$, $Al^{3+}$, $K^+$, and $Na^+$ were all highly positively correlated with the concentration of $NO_3^-$ (Likens et al., 1969, 1970). These data show the quantitative importance of nitrification as a major controlling factor in determining the quantity and quality of dissolved substances flushed from the deforested watershed.

*Effect of Transferring Living to Dead Biomass.* A final consideration is the short-term effect of clear-cutting, without product removal, on the living to dead biomass ratio and on the export of nutrients. Large amounts of nutrients are stored in the living biomass within the ecosystem. Based on our analysis of our 55-yr-old aggrading forest, we estimate that, prior to cutting, living biomass on W2 contained the following quantities of nutrients (in kilograms per hectare): N, 532; Ca, 484; Mg, 49; P, 87; K, 218; S, 59 (Likens et al., 1977, p. 101). One might think that these nutrients were the major source of nutrients flushed from the ecosystem during the devegetation period. Although there is no doubt that these nutrients would play an important role in the subsequent recovery of the deforested ecosystem, they apparently played only a minor role in the behavior of dissolved substances during the 3-yr period.

Two lines of evidence support this conclusion: (1) Nutrient losses from the ecosystem are adequately explained by diminishment of soil organic matter and exchangeable nutrients (Dominski, 1971), and (2) nutrient relationships during devegetation would most likely be affected by addition of easily decomposable substrates resulting from cutting. Such additions were small when compared to the nutrient reservoir in the soil at the time of cutting. The easily decomposable tissues are leaves and fine twigs and roots; branches, stems, and larger roots decay over longer periods of time (Spaulding and Hansbrough, 1944). The forest on W2 was cut *after* normal leaf fall, hence leaves were already part of the forest floor. Whittaker et al. (1978) estimate 3.5, 0.3, and 1.0-kg/ha of N, P, and K in the standing crop of current twigs. We have no direct estimate for small roots, but if we assume a value similar to that of twigs the total amount is not great. Assuming all nutrients in this mass were quickly mineralized during the first warm season after cutting, the effect on nutrient behavior would be minor.

Wood and bark of branches, stems, and roots decay much more slowly. Moreover, woody tissues have a relatively low nutrient content (Likens and Bormann, 1970). Nutrients released from these tissues would be significant during the later part of the recovery period.

*Integration of Processes.* A hypothetical set of interactions affecting dissolved substance concentrations during the devegetation period is summarized in Figure 3-5. Although large quantities of living plant biomass were shifted to dead biomass by cutting, only a relatively small quantity was decomposed in the first 2 or 3 years after deforestation. Activities of decomposers, in terms of carbon release, were accelerated in comparison with the forested system owing to more favorable conditions of soil temperature and moisture and the greater concentration of nutrients in the soil solution. As a result of the activities of decomposers, quantities of anions and cations were released into the soil solution. These ions come both from atoms incorporated in molecules making up the biomass (mostly forest floor) and from exchangeable ions released by

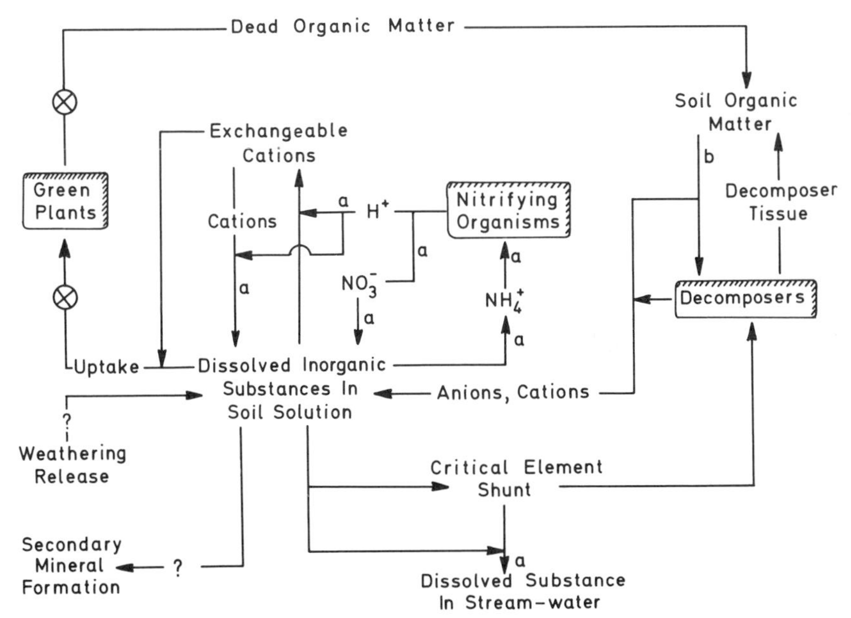


**Figure 3-5.** Hypothetical set of interactions affecting concentrations of inorganic dissolved substances in the soil solution and in stream water draining Watershed 2 during the 3-yr devegetation period. (⊗) Block placed on an ecosystem pathway by the experimental design (*a*) transfer increased over that expected in vegetated system for the entire 3-yr period; (*b*) movement of carbon to decomposers (at first greater than a vegetated system and then less). The critical element shunt operates as follows: at first, when C/N or element/P ratios of decaying organic matter are high, dissolved nutrients are taken up by microorganisms, as these ratios decrease nutrient export increases in relation to the drop in nutrient demand by microorganisms. We propose that action of the shunt may account for highest nutrient concentrations in drainage water during the second year of devegetation.

decomposition of exchange surfaces. Ammonium cations, a large portion of which would normally be taken up by green plants, were oxidized by nitrifying microorganisms to nitrate, with the production of two hydrogen ions. Some of these hydrogen ions are exchanged for basic cations on exchange sites, resulting in the release of these basic cations to the soil solution. Other hydrogen ions remained in the soil solution and subsequently lowered the pH of stream water from 5.1 to 4.3 in 1967–68 (Likens et al., 1970).

The increase in nitrification is of particular interest because the response time was rapid (i.e., within a few weeks of the onset of the growing season as judged by stream-water concentrations, Likens et al., 1970) and because its initial response was not dependent on an increased rate of decomposition. As previously mentioned, ammonium normally used by green plants was nitrified in the deforested ecosystem. Nitrification may have enhanced decomposition rates by producing $H^+$ ions, which in turn exchanged for cations on exchange sites, enriching the nutrient content of the soil solution and making potentially limiting cations available to fungi and bacteria. If this were so, nitrification and decomposition could positively reinforce each other over the short term. This interaction would ultimately be limited by other components within the system, such as quantities of exchangeable cations and the amount and quality of decomposable substrate.

The quantity and quality of decomposable substrate exercises important regulatory control over the biogeochemical behavior of the ecosystem. The most easily decomposable substances in fresh litter (such as sugars, starches, simple proteins, and crude proteins) are acted upon most rapidly and vigorously. As these substances are diminished, more resistant substances such as cellulose, lignins, fats, waxes, polysaccharides, and polyuronides from dead microbial tissue, or, in general, humic compounds, become the principle substrate for decomposers, and rates of carbon release diminish (Brady, 1974). Most decomposition occurs in the forest floor, but substantial losses may occur at lower depths (Dominski, 1971).

Some dissolved substances are carried directly out of the ecosystem, but probably some are immediately taken up by the decomposers and do not appear in stream water draining the ecosystem. However, it seems likely that, as the microbial demand decreases owing to a lowering of C/N ratios and/or element/phosphorus ratios, dissolved nutrients that previously moved to decomposing organisms would be released in stream water. We call this the "critical element ratio shunt." During the first part of the devegetation period the shunt moves elements toward decomposers, but during the later part of the period when critical element ratios are more closely approached the shunt is opened, and additional dissolved substances are lost in stream water draining the ecosystem. This could explain higher dissolved substance losses in stream water during the second year.

## Erosion of Particulate-Matter

Clear-cutting and enforced devegetation had a pronounced but delayed effect on the normally low rate of particulate matter export characteristic of aggrading forest ecosystems (Bormann et al., 1974). Output of particulate matter from the deforested ecosystem (W2) prior to 10 October 1967, which included two growing seasons after cutting, was about three times that of the aggrading ecosystem. During the period 10 October 1967 to 13 August 1969, which included the last growing season of the devegetated period plus the first 2 months of the regrowth period, output from W2 was 16 times greater than W6, our forested reference watershed (Figure 3-6).

These data illustrate a very important point: despite removal of all living vegetation and suppression of regrowth for three growing seasons, the capacity of the ecosystem to resist the force of erosion remained comparatively high for more than 22 months. Thereafter, particulate matter losses increased rapidly until new vegetation was firmly reestablished.

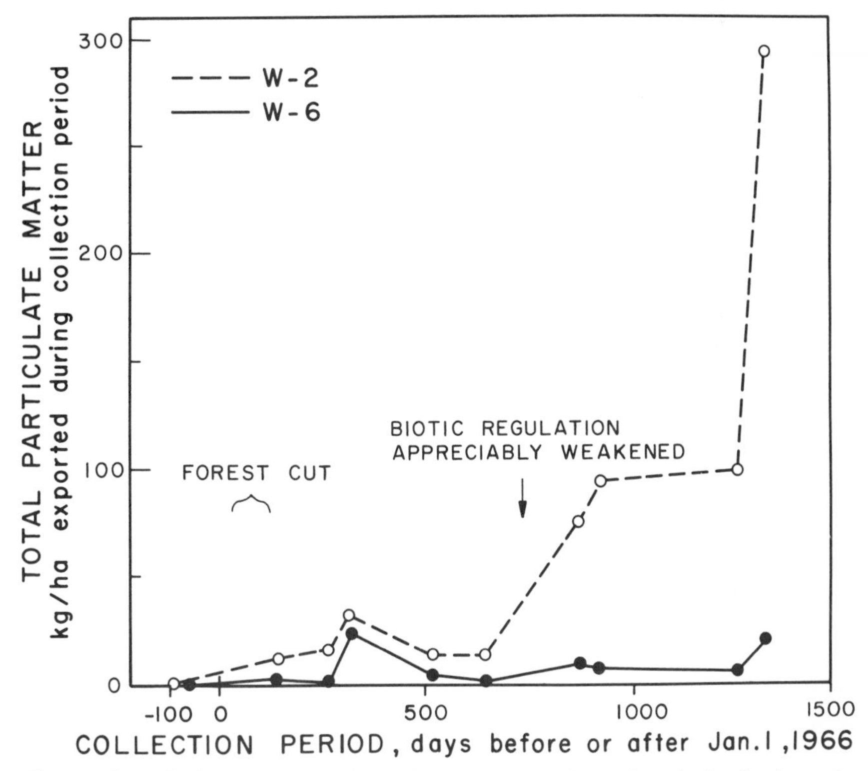


**Figure 3-6.** Sediments, organic and inorganic, collected periodically from the ponding basins of the devegetated ecosystem, Watershed 2 (○--○) and from a forested ecosystem, Watershed 6 (●—●). Collections cover the period 1 June 1965 to 13 August 1969 (from Bormann et al., 1974).

Two factors contribute to this pattern of particulate-matter losses. These are changes in hydrology and erodibility resulting from clear-cutting and enforced devegetation. This is best understood by considering the relationship of particulate-matter concentration to flow rate, as shown in the composite curve for the forested ecosystem (Figure 2-7).

The form of the curve is a function of (1) the increasing capacity of moving water to do the work of erosion and transportation as velocity increases and (2) the erodibility of the ecosystem. Velocity ($V$), often positively correlated with discharge, is highly correlated with sediment discharge (Gilbert, 1914; Leopold et al., 1964). As an approximation, Kittredge (1948) states that the erosive power of water is proportional to $V^2$.

Deforestation can thus alter the pattern of particulate-matter export in two ways: (1) by changing hydrologic characteristics of the ecosystem through increases in the velocity of storm peak streamflow and in the amount of streamflow at higher velocities and (2) through increases in erodibility of the ecosystem (Table 3-3).

Deforestation markedly increased erodibility, although the full impact occurred only after a considerable lag period. This effect was easily seen, despite the fact that in the forested ecosystem erodibility is not a constant: it changes seasonally and during individual storms—and the ecosystem may be more erodible on the rising limb of the hydrograph than on the falling limb (Leopold et al., 1964; Striffler, 1964; Bormann et al., 1969, 1974; Fisher, 1970).

During the first portion of the devegetation period, January 1966 to October 1967, changes in hydrology accounted for an increase in

**Table 3-3.** Estimates of Increased Output of Particulate Matter due to Changes in Hydrology and Erodibility[a]

| | 1 June 1965 to 9 October 1967 | 9 October 1967 to 13 August 1969 |
|---|---|---|
| Amount expected if forest had not been cut | 551 | 662 |
| Increase due to changes in hydrology | 249 | 1235[b] |
| Increase due to changes in erodibility[c] | 691 | 6937 |
| Total amount measured behind weir | 1491 | 8834 |

*[a]Based on particulate matter collected from the ponding basin behind the weir of W2; all data in kilograms per hectare (from Bormann et al., 1974).*
*[b]There was an exceptionally large storm peak during this period.*
*[c]Slightly different than previously reported values (Bormann et al., 1969) owing to correction of a transcription error.*

particulate-matter output by a factor of 0.5 over that expected if W2 had remained forested, while changes in erodibility accounted for an increase by a factor of 1.3. Between October 1967 and August 1969, which includes part of the first year of revegetation, hydrologic changes accounted for an increase of about 1.9 times, while changes in erodibility contributed to an increase 10.4 times the amount expected if cutting had not occurred (Figure 3-7).

These data indicate that mechanisms governing the erodibility of the aggrading northern hardwood forest ecosystem continue to function after deforestation at or near the level of the forested ecosystem for about 2 years, including two growing seasons. The ability of these erosion-controlling mechanisms to operate for an extended period when primary productivity and nutrient uptake have been disrupted is an important component of the stability of the northern hardwood forest ecosystem.

The delayed response of the ecosystem to erosion confirms the findings of foresters (Hoover, 1944; Reinhart et al., 1963; Lull and Reinhart, 1967, 1972) that the removal of forest cover in well-developed eastern forests has little effect on sediment yield as long as the forest floor is intact.

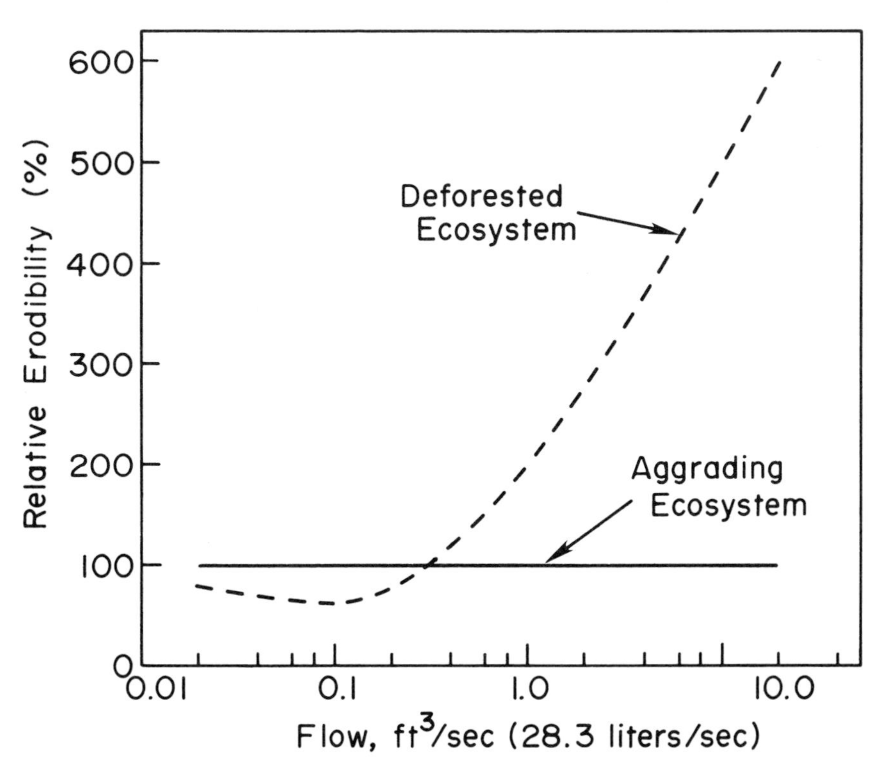


**Figure 3-7.** The increase in erodibility of Watershed 2 in relation to Watershed 6 at the treatment period 9 October 1967 to 13 August 1969 (after Bormann et al., 1974).

Apparently, erodibility was controlled to a large extent by residual dead organic matter in the ecosystem at the beginning of the deforestation period. About 2 years were required before this control was weakened substantially by decomposition and mechanical factors. Loss of less than a fourth of the organic matter in the forest floor is associated with a major increase in ecosystem erodibility. What is not known, however, is the amount and rate of decomposition of the mass of dead roots and rootlets in the stream banks lining the stream channel and in the organic-debris dams within the channel. It seems likely that decomposition of these materials would have contributed substantially to the increased quantities of particulate matter flushed from the ecosystem.

## Overall Effects of Enforced Devegetation on Ecosystem Dynamics

Destruction of vegetation and suppression of regrowth essentially eliminated functions of green plants, which are a normal part of the forested ecosystem. Active and passive processes such as uptake of water and nutrients, transpiration, canopy absorption and reflection of radiant energy, primary production, competition with nitrifying microorganisms for $NH_4^+$, allelopathic effects on nitrification, and annual litter transfer were essentially eliminated for 3 years by the enforced devegetation.

The net result of interactions triggered by deforestation and maintained by herbicide application is pictured in Figure 3-8. The interactions are complicated and involve major shifts from the highly predictable conditions found in the undisturbed aggrading forest. The following effects are of particular interest.

1. Storm peaks were accentuated, and annual streamflow was increased by about 30%, mostly during the growing season. The control that aggrading forests have over summer hydrology was essentially lost. This occurred because the uptake of water and transpiration by green plants was eliminated and interception was reduced (Pierce et al., 1970).
2. Concentrations of most dissolved nutrients in stream water increased severalfold. The highly predictable relationship between nutrient concentrations and flow rate of streams draining aggrading ecosystems are severely disrupted. Primary factors involved in this increase were soil moisture and soil temperature conditions favorable to an increase in the rate of decomposition and an array of factors favorable to a rapid increase in the rate of nitrification. It seems likely that initially the processes of decomposition and nitrification acted in a positive-feedback cycle, with each factor reinforcing the other. Concentrations of dissolved substances in stream water rose rapidly during the first growing season after cutting, but the highest concentrations were recorded during the second growing season.
3. The export of dissolved substances in stream water increased rapidly

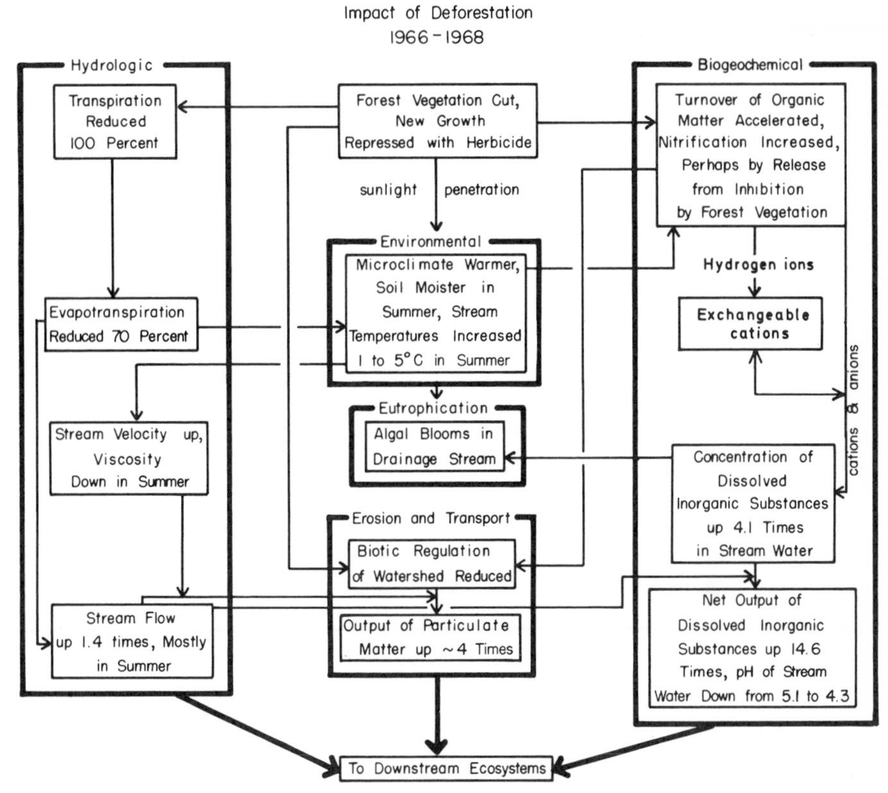


**Figure 3-8.** The impact of deforestation on the hydrology, microclimate, biogeochemistry, and the erosion and transport of particulate matter during the first 2 water-years after cutting (Likens and Bormann, 1972a).

after deforestation, and during the 3-yr period the total net losses were about eight times those of a forested ecosystem. The highly developed capacity of the undisturbed forest to hold nutrients within the ecosystem and to recycle them was severely diminished. This response resulted from the interaction between Items 1 and 2 above.

4. The "excess" nutrients exported from the devegetated ecosystem were principally withdrawn from those incorporated in soil organic matter and held on exchange surfaces. The upper layer of the soil, i.e., the forest floor, was the source of most of the dissolved substances. This is in contrast to the aggrading forest, where the forest floor accumulates biomass and nutrients.
5. Toward the end of the devegetation period, erodibility of the ecosystem as measured by losses of particulate matter increased about 10 times over that of the aggrading ecosystem. The effectiveness of running water to remove both organic and inorganic particulate matter was thereby increased. The weakening of biotic regulation of erosion was related to the suppression of growth and

to organic decomposition processes proceeding in the absence of annual litter input. However, biotic regulation was relatively unimpaired during the first two growing seasons after cutting; thereafter, it increased exponentially.

## RELATIONSHIP OF THE DEFORESTATION EXPERIMENT TO COMMERCIAL CLEAR-CUTTING

Are conclusions resulting from the clear-cutting and enforced-devegetation experiment applicable to clear-cuts where revegetation occurs quickly? We think they are, and it is probable that the same pattern of changes shown in Figures 3-5 and 3-8 occurs in commercially clear-cut northern hardwood forests, even though regrowth obscures some of the responses seen in the devegetated ecosystem. We base this conclusion on several lines of reasoning.

Commercial clear-cutting initially disrupts, to an extent equivalent to that seen in W2, active and passive processes of green plants. These processes are fundamental to the web of interactions shown in Figures 3-5 and 3-8. Since time is required for primary production to return to previous levels, we expect the first year after commercial cutting to be most similar to that seen in W2; i.e., green plant production will be relatively low and decomposition will be high. Thereafter, we would expect the response of the two systems to diverge as production increases in one while the other remains devegetated.

We would expect the hydrologic response of the commercially clear-cut ecosystem to be very similar to that of the devegetated system during the first year, since increased streamflow in humid forest ecosystems is closely related to the areal extent of cutting (Hibbert, 1967).

We examined the stream-water chemistry of six commercially clear-cut northern hardwood forests scattered throughout New Hampshire but occurring under conditions similar to those at Hubbard Brook. We found trends in stream-water chemistry similar to those found in W2 for a 2-yr period after cutting (Pierce et al., 1972). These findings (Table 5-2) showed (1) a general increase in concentrations of dissolved substances when compared to a nearby stream draining a forested watershed, (2) a marked increase in nitrate output, and (3) a tendency for maximum concentrations in stream water to occur during either the first or second year after cutting.

Export of dissolved substances from W2 exceeds that of the commercially clear-cut ecosystems. On the other hand, losses due to removal of wood products did not occur in W2 but were substantial in commercially-cut stands. In fact, total losses of calcium and nitrogen from the commercial clear-cut at Gale River equaled or exceeded those from W2 (Table 3-4). This is of importance over the long term, when nutrients locked in slash once again become available. When that happens, W2 could have a significant advantage over a commercially clear-cut stand.

**Table 3-4.** Comparative Losses of Dissolved Calcium and Nitrogen in Stream Water during the First 2 Years after Clear-Cutting and Losses Incurred by the Removal of Wood Products[a]

| | Hubbard Brook Experimental Clear-Cut[b] | | Gale River Commercial Clear-Cut | |
|---|---|---|---|---|
| | Calcium | Nitrogen | Calcium | Nitrogen |
| First year | 75 | 96 | 41[c] | 38[c] |
| Second year | 90 | 140 | 48[c] | 57[c] |
| Removed in wood products | 0 | 0 | 221 | 144 |
| Total removed | 165 | 236 | 310 | 239 |

[a] *Values are in kilograms per hectare.*
[b] *From Likens et al., 1970.*
[c] *From Pierce et al., 1972.*

Studies of the forest floor after commercial clear-cutting show a progressive net decrease in both weight and depth for about 10 to 20 years after cutting, despite growth (Morey, 1942; Trimble and Lull, 1956; Hart, 1961; Covington, 1976). This parallels the initial behavior of W2 and indicates a period of time after cutting when decomposition in the forest floor exceeds litter input in both ecosystems.

We would expect the erosion of particulate matter from any undisturbed surfaces in the commercially cut ecosystem to parallel that in W2 during the first year. Thereafter, losses of particulate matter would be lower owing to the restoration of biotic regulation of erodibility. However, overall, erosion of the commercially cut ecosystem would tend to rise above that of W2 because of mechanical disruption of surfaces by road building and harvesting techniques.

Thus, the clear-cutting and enforced-devegetation experiment provides a good qualitative model (Figures 3-5 and 3-8) of the interactions triggered by destruction of the dominant vegetation of the northern hardwood ecosystem. The quantitative aspects of the models, when applied to a commercial clear-cut, would vary in relation to the rapidity with which active and passive plant processes were restored after cutting.

Although the model emphasizes loss of biotic regulation typical of the aggrading ecosystem after destruction of vegetation, the biogeochemical responses shown in Figures 3-5 and 3-8 are part of another kind of biotic regulation that leads to the rapid recovery of the disturbed ecosystem. This type of regulation will be dealt with in Chapter 5.

## Summary

1. An experiment was designed to examine the effects of vegetation destruction on a variety of biotic and abiotic processes within the ecosystem.

2. Destruction of the vegetation and prevention of regrowth had the following results:
   a. Solar energy flow was markedly changed because of removal of the reflective surface of the canopy and the elimination of the use of energy in transpiration. Changes in energy flow had the dual effect of increasing summertime soil temperatures while increasing soil moisture. Lack of transpirational drawdown caused increased moisture in the soil.
   b. Vegetative control over summertime hydrology was markedly diminished. This resulted in increased export of stream water and increased storm-peak flow rates.
   c. The highly predictable relationships between nutrient concentrations in stream water and flow rate, typical of the aggrading ecosystem, are rapidly lost after vegetative destruction. This results in increased nutrient concentrations and markedly increased losses of nutrients from the ecosystem.
   d. Increases in the rates of decomposition and nitrification play an important role in the nutrient relationships of the disturbed ecosystem.
   e. Most nutrients lost from the ecosystem originate in the forest floor and are released by decomposition–nitrification interactions. For a time these processes may have a positive-feedback relationship.
   f. Maximum concentrations and losses of nutrients occurred in the second year after devegetation. This may result because high C/N or element/P ratios in dead organic matter lead to increased utilization of dissolved nutrients by microorganisms during the first year. As ratios are lowered, dissolved nutrient concentrations in stream water increase.
   g. The process of nitrification, which is partially inhibited in the aggrading ecosystem, is accelerated by about an order of magnitude in the devegetated system.
   h. Clear-cutting and enforced devegetation have the pronounced but delayed effect of increasing the very low rate of particulate matter export characteristic of aggrading forests. Increases in export are due primarily to increases in erodibility rather than increases in storm peaks resulting from deforestation.
3. Two models are presented showing the interactions that affect dissolved substance concentrations in the soil of the devegetated ecosystem and the impact of deforestation on the hydrology, microenvironment, biogeochemistry, erosion, and transport of particulate matter in the ecosystem.
4. Evidence is presented showing that models developed for the ecosystem that was clear-cut and held bare are applicable to commercially clear-cut forests.
5. It is proposed that biogeochemical changes resulting from destruction of the canopy contribute to the rapid recovery of the ecosystem.

CHAPTER 4

# Development of Vegetation After Clear-Cutting: Species Strategies and Plant Community Dynamics

The focus of this chapter is the regrowth of vegetation during the Reorganization and Aggradation Phases. Our emphasis will be on plant community dynamics and species strategies, but it must not be forgotten that ecosystem development involves an array of organisms—plants, animals, and microorganisms. Parasites, predators, symbionts, and decomposers may speed or direct changes in plant populations, or they themselves may decline or disappear as a result of changes in plant-community dynamics. Integrating our knowledge of these relationships represents a major challenge for future ecosystem studies.

## WHAT IS SECONDARY SUCCESSION?

Secondary succession has often been thought of as a fairly orderly process of development that follows the abandonment of cultivated fields. Relays of higher plants, many of which were not present in the cultivated field, follow each other; a sequence of revegetation stages occurs, and, ultimately, "climax" vegetation is established (Dansereau, 1957). This pattern is not true for secondary succession after clear-cutting of northern hardwoods (when the procedure is one of careful cutting, product removal, little soil damage, and immediate revegetation). Rather, the pattern after clear-cutting more closely approaches that proposed by Egler (1954) for abandoned pastures, where the initial floristic composition at the time of abandonment accounts for 95% of later development. Although the relative importance of individual species will change as their life strategies unfold and as shifting stages of dominance occur in the ecosystem (Harper and Ogden, 1970), the great bulk of the species that will participate in the changes that follow clear-cutting are present, in active or dormant form, in the intact forest. These individuals are released by cutting and begin to grow more rapidly while still other species enter the site within a few years after cutting. Not only are all of these species present shortly after cutting but so too are individuals that will compose the majority of dominant trees for more than a century if there is no serious exogenous disturbance.

The divergence in patterns of secondary succession between formerly cultivated fields and cutover forest is not unexpected, given the nature of the perturbing forces. When a forest is converted to arable land, not only is the original vegetation diminished, but the site is subjected to the continuing perturbations of plowing, cultivation, pesticides, and fertilizers. The net effect is a progressive reduction in the pool of species present in the original ecosystem. Thus, in the southeastern United States, for example, one does not find significant numbers of oaks, hickories, pines, or dozens of other forest species, in cultivated fields. These species gradually reenter the site after cultivation ceases and a variety of other conditions are met, such as good seed years and appropriate environments for seed dissemination and establishment; they may then reenter the ecosystem in a fairly predictable sequence (Oosting, 1956).

Clear-cutting of northern hardwoods, on the other hand, represents a *single* severe perturbation followed by growth of species already present in the ecosystem or that enter it shortly after cutting. Ecosystem development is marked by a sequence of dominant trees similar to that observed in some old-field developmental sequences. However, for a century or more, dominants are largely drawn from the pool of individuals already established at the time of cutting. These dominants fit into the developmental continuum according to their specific strategies. Eventually, new individuals (but not necessarily new species) enter the ecosystem and make their way into the dominant canopy, but this mode of canopy replacement becomes important only after a considerable length of time.

## REPRODUCTIVE AND GROWTH STRATEGIES RESPONSIVE TO PERTURBATIONS THAT OPEN THE FOREST CANOPY

At the beginning of our hypothetical developmental sequence (the limits of which have been discussed in Chapter 1), the northern hardwood forest contains a wide array of higher plant species. At various points along the developmental continuum, activities of some species are accelerated, while others are decelerated, suppressed, or eliminated in relation to the degree to which their niche specifications are met. Succession, or the continuum of change in vegetation composition and dominance, is thus the result of constantly changing conditions of selection acting on a large and varied pool of species strategies.

The selection–strategy interaction involves not only changing selection forces but many life strategies, each with a wide range of niche dimensions. To make some sense of this complexity, we first approach this interaction from a restricted point of view by looking at the capacity of species to fill space in the dominant (upper) layer of the ecosystem when openings in that layer are created by some kind of nonpersistent disturbance ranging from a single treefall to clear-cutting. We set forth a variety of reproductive and growth strategies by which species of the northern hardwood forest may act to fill holes created in the canopy by various disturbances (Table 4-1).

These strategies are of two general kinds: *outgrowth*, in which existing individuals (Trimble and Seegrist, 1973) or clones expand laterally to fill in newly created space, and *upgrowth*, in which individuals or clones grow upward to form a new upper or locally dominant canopy.

*Upgrowth* may be from individuals established in the intact forest prior to disturbance (advance regeneration). This would include stump and root

**Table 4-1.** Growth and Seedling Strategies that Function to Fill in the Dominant Layer in Disturbed Areas of Northern Hardwood Ecosystems

I. *Outgrowth:* Mostly lateral or horizontal expansion of established individuals into space temporarily freed from intense competition
   A. Clonal expansion of herbs and shrubs, e.g., hay-scented fern, hobblebush, raspberry
   B. Encroachment into an opening by crowns and roots of surrounding trees
II. *Upgrowth:* Height growth that forms a rising overstory canopy beginning at or near ground level
   A. Advance regeneration: Stems growing on the site
      1. Stump and root sprouts, e.g., beech, striped maple, mountain maple
      2. Seedlings and saplings, e.g., beech, sugar maple, ash, yellow birch
   B. New individuals: Stems not previously growing on the site
      1. Stump and root sprouts from injured individuals, e.g., red maple, aspen
      2. New germinants
         a. Stored or buried seeds accumulated on the site, e.g., pin cherry, raspberry, elderberry
         b. Current seed crop
            1. Heavy-seeded species, which may produce abundant seedlings if a large crop of seed was produced just prior to disturbance or if there are nearby seed sources; e.g., beech, sugar maple, ash
            2. Light-seeded or fugitive species (Marks, 1974), which may produce abundant seedling populations from seed produced on the site prior to disturbance or from seed introduced into the site from considerable distance and are typified by abundant and frequent production of light, highly mobile, wind-transported seed; e.g., aspen, birch

sprouts and growth of established seedlings or saplings having various levels of tolerance to shade (Table 4-2). It is important to note that a variety of nuances exist within the "advance regeneration" strategy. Shade-tolerant species such as beech, sugar maple, and red spruce are often able to maintain, in fairly deep shade, large populations of seedlings whose ages range over several decades (Forcier, 1973). Other less tolerant species like white ash and striped maple have somewhat less aggressive advance-regeneration strategies. Usually, their seedlings are found in somewhat better lighted locations in smaller numbers, and the lifespan of seedlings that are not released by disturbance is much shorter than their more tolerant colleagues. At the lowest end of the spectrum are species like yellow birch, whose seedlings germinate and become established within the closed forest but whose longevity is generally less than 1 year (Forcier, 1973).

*Upgrowth* after disturbance also may result from new individuals or stems not actively growing in the undisturbed site. This may come from dormant buds, their growth stimulated by injury to parent plants, or from germination and establishment of seedlings. Germination of buried seeds is one of the most important revegetation mechanisms in the northern hardwood ecosystem, and this unique adaptive strategy will be considered in more detail in the next section.

*Upgrowth* may result from growth of seedlings of the current seed crop

**Table 4-2.** Shade-Tolerance Ratings of Northern Hardwood Trees[a]

| | | |
|---|---|---|
| *Tolerant* (able to survive in deep shade): | | |
| beech | sugar maple | red spruce |
| balsam fir | hemlock | |
| *Intolerant* (generally unable to survive very long in deep shade): | | |
| white birch | trembling aspen | gray birch |
| pin cherry | bigtooth aspen | |
| *Intermediate:* | | |
| yellow birch | red maple | striped maple |
| mountain maple | white ash | |

[a] *Tolerance ratings, although vague in describing the response of a species to specific environmental conditions, are useful and in a sense represent a consensus of many workers' perceptions of a species' ability to grow under a variety of conditions. As a consequence, tolerance ratings represent a good starting point for discussing the relationship of a species to environmental factors. Most frequently, tolerance is expressed on a scale and interpreted as the ability of a species to survive and grow in their own shade and that of associated species. Baker (1949) points out the lack of precision of tolerance scales but defends their usefulness. He and Horn (1971) also recognize that the tolerance rating of a species may change under different overstories or with different site conditions. Nevertheless, the concept has definite utility in discussing secondary succession in northern hardwood ecosystems. In this regard, we follow workers of the U.S. Forest Service (Leak, Solomon, and Filip, 1969; Gilbert and Jensen, 1958).*

produced on or off the site. Salisbury (1942) in his superb treatise *The Reproductive Capacity of Plants* pointed out that species with light, highly mobile seeds often invade highly disturbed situations while species with heavier seeds, often animal disseminated, usually enter the ecosystem at a later stage of development. In clear-cut areas, upgrowth by light-seeded species such as aspen and white birch (which are transported to the site by wind) may play an important role in revegetation, depending upon the degree of soil disturbance, nearness of seed sources, and other stochastic factors. *Upgrowth* by heavy-seeded species transported to the clear-cut site after disturbance is usually less important, but in the long run it is movement of heavy seeds by wind, mammals, or birds that is responsible for maintaining the geographic distribution of the many heavy-seeded species which make up a sizable proportion of the flora of the northern hardwood ecosystem.

It is important to note that few, if any, species are dependent on a single reproductive strategy; most species employ several strategies, with the importance of each shifting in relation to the nature of the disturbance and other circumstances. An analysis of the most vigorous trees of each species in a 6-yr-old stand grown up after the clear-cutting of a 60-yr-old northern hardwood forest gave an interesting insight into the relative importance of advance regeneration and new germinants in providing future dominants of the site. We established 50 sample points within the stand and harvested the largest individual of each species encountered within a 3-m radius of each point. Generally, intolerant species were the tallest, followed by species of intermediate tolerance, and tolerant species were shortest. We measured the age of each individual to determine whether it was already established on the forest floor at the time of cutting or was a new germinant (Table 4-3). Rather

**Table 4-3.** Percentage of the Largest Saplings, by Species, that Came from Advance Regeneration and New Germinants in the Sixth Growing Season after Clear-Cutting of a 60-Yr-Old Northern Hardwood Forest[a]

| | | Advance Regeneration (%) | | New Individuals (%) | | Percentage for Each Species | |
|---|---|---|---|---|---|---|---|
| Species | Total no. of Trees | Sprouts | Seedlings | Sprouts | New Germinants | Advance Regeneration | New Individuals |
| Beech | 49 | 3 | 39 | 7 | 0 | 86 | 14 |
| Mountain maple | 21 | 0 | 14 | 0 | 7 | 67 | 33 |
| Sugar maple | 50 | 4 | 17 | 8 | 21 | 42 | 58 |
| Striped maple | 50 | 1 | 19 | 1 | 29 | 40 | 60 |
| White ash | 49 | 0 | 11 | 3 | 35 | 22 | 78 |
| Yellow birch | 50 | 0 | 4 | 1 | 45 | 4 | 92 |
| Pin cherry | 50 | 0 | 0 | 0 | 50 | 0 | 100 |
| Trembling aspen | 15 | 0 | 0 | 0 | 15 | 0 | 100 |

[a]*Based on 50 circular plots 6-m in diameter; the largest individual of each species was harvested in each plot; data are limited to species presented on 15 or more plots.*

surprisingly, for six of the eight species the majority of the largest trees were new germinants. Based on these data, relative dependence on the advance-regeneration strategy was beech > mountain maple > sugar maple and striped maple > ash and yellow birch. Pin cherry probably came from buried seed, and aspen from seeds blown onto the site (Figure 4-1).

## THE BURIED-SEED STRATEGY

The buried-seed strategy is of particular significance in revegetation after clear-cutting. In fact, behavioral patterns of species using this strategy suggest that it may have evolved specifically in response to the highly specialized conditions that follow serious disturbance of a previously intact ecosystem. In terms of system dynamics, the strategy is not simply exploitation of an injured ecosystem but is also part of an elaborate feedback mechanism that acts rapidly to limit temporarily accelerated nutrient export and to return the system to a condition of greater stability. The buried-seed strategy involves: (1) storage of large quantities of dormant seeds in the soil of the intact forest, even though vegetative individuals may be wholly absent; (2) triggering of seed germination by an environmental signal(s) associated with disturbance of the canopy; (3) rapid height growth associated with or dependent upon temporary nutrient enrichment of the soil, which results from the accelerated heterotrophism that accompanies disturbance (Chapter 3); (4) the relatively short period of time before abundant seed production occurs; and (5) a usually short lifespan of the individuals.

The occurrence of buried seeds and their germination after disturbance has been known for a long time (Marks, 1974). Recently, Bicknell (Table 4-4) and Marquis (1975) have estimated the number of buried seeds in northern hardwood forests of various ages. Literally millions of viable seeds per hectare are buried in the soil under many northern hardwood forests. It is important to note, however, that not all buried seeds are long-lived. Seeds of some species, such as white ash (Clark, 1962; Leak, 1963) and yellow and white birch, have a very limited longevity, and, in fact, these species seem to rely on two strategies: buried seeds and/or direct seeding just before or after disturbance. On the other hand, the viability of seeds of species like raspberry, pin cherry, and perhaps elderberry extends into many decades and perhaps to more than a century (Olmsted and Curtis, 1947; Marks, 1974; Whitney, 1978).

Marks (1974) has addressed the question of how enormous numbers of heavy seeds, such as those of pin cherry, come to reside in soil even though mature trees are no longer present. He proposes two cycles of pin cherry dissemination: a low-density cycle that leads to widespread dissemination of the species and a high-density cycle that leads to massive occurrence of individuals on local sites (Figure 4-2). In today's forest, pin

**Figure 4-1.** A 5-yr-old stand of pin cherry in a small cleared area of the Hubbard Brook Experimental Forest. The stand originated from germination of buried seeds after the 50-yr-old forest was cut. Density is about 10 stems/m$^2$, and trees are about 4-m tall.

**Table 4-4.** Viable Buried Seeds in Northern Hardwood Stands of Various Ages[a]

| | | Number of Seeds (millions/ha) | | | | | |
|---|---|---|---|---|---|---|---|
| | Estimated Longevity of Dormant Seeds | Age of Stand in Years | | | | | |
| Species | (yr)[b] | 5 | 15 | 20 | 30 | 60 | >100 |
| Red maple | 3–15 | — | — | — | 0.1 | 0.2 | — |
| Yellow or white birch | 3–15 | 0.6 | 0.6 | 1.2 | 2.4 | 1.5 | 0.8 |
| *Erigeron sp.* | < 3 | 9.3 | 0.1 | — | — | — | — |
| Pin cherry | >15 | 0.2 | — | 0.2 | — | 0.2 | 0.1 |
| Raspberry or blackberry | >15 | 0.1 | 4.9 | 1.2 | 1.3 | 2.6 | 3.5 |
| Elderberry spp. | >15 | 1.1 | 0.2 | 1.0 | — | 0.2 | 0.8 |
| Goldenrod spp. | < 3 | — | 0.3 | 0.1 | — | — | — |
| Violet spp. | >15 | — | 0.3 | — | 0.1 | — | — |
| Grasses spp. | >15 | — | 0.3 | — | 0.4 | 0.1 | — |
| Sedges spp. | >15 | 3.8 | 6.4 | 0.1 | 2.0 | 0.1 | — |

[a] *Determined by germination of seeds contained in twenty 10×10-cm blocks cut out of the forest floor. Germination per block was very variable, ranging from zero to several orders of magnitude greater than the mean (S. Bicknell, unpublished data).*
[b] *Adapted from Harrington (1972).*

cherry is common on cutover and burned sites. It has a lifespan of about 30 years in the Northeast, but during this time it can produce 15 to 20 million fruits per hectare in well-stocked stands. Many birds are attracted to these sites, where they consume the fruits, digest the pericarps, and excrete the pits containing the seeds (Thompson and Willson, 1978). The ability of birds to disseminate pin cherry seeds seems great (Marks, 1974). McAtee (1910) lists 39 species, Olmsted and Curtis (1947) 33 species, and Martin et al. (1951) 47 species of birds that feed on *Prunus*. In this way pin cherry seeds are

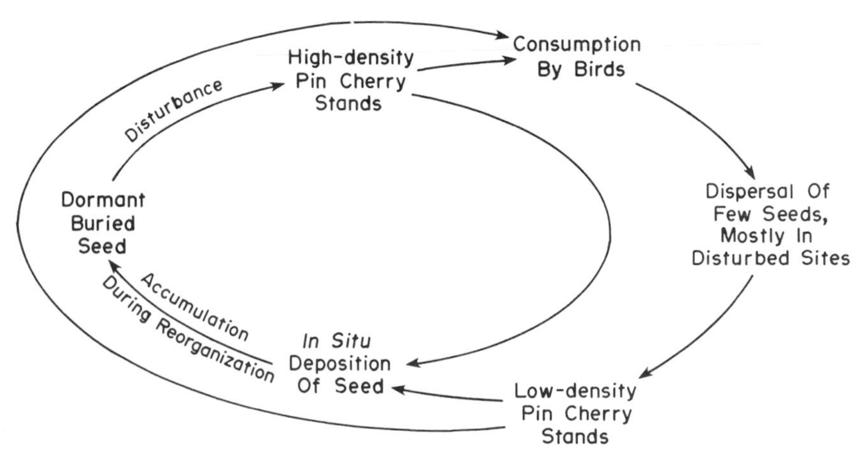


**Figure 4-2.** Representation of high- and low-density pin cherry cycles (modified from Marks, 1974).

moved to other disturbed sites frequented by the foraging birds. This mechanism serves to introduce pin cherry into newly disturbed sites not recently occupied by the species and thus contributes to the buried-seed pool. With the next cycle of disturbance, scattered individuals of pin cherry occur and produce large numbers of seeds, many of which become buried in the soil and remain dormant. Another cycle of disturbance may then lead to a pin cherry stand of high density and still-greater numbers of buried seeds. Marks' hypothesis probably applies to other buried-seed species, such as raspberry and elderberry.

In many second-growth northern hardwood forests, where pin cherry trees are absent or are a very minor component, hundreds of thousands of dormant pin cherry seeds may be found in a hectare of soil (Table 4-4; Olmsted and Curtis, 1947; Marks, 1974). The size of the seed pool suggests that most dormant seeds are produced in the site early in the revegetation sequence rather than being transported to the site in small increments by animals.

The mechanism(s) by which germination of buried seeds is synchronized with disturbance of the ecosystem represents one of the paramount control features in ecosystem dynamics; in a sense, it initiates a homeostatic response held in reserve for just such an occasion. Disturbance (cutting, burning, or windthrow) triggers the germination of large quantities of buried seeds. Simply moving a volume of forest floor to the greenhouse will promote germination of many buried seeds. However, the triggering mechanisms underlying germination in response to disturbance are far from clear.

Marks (1974) has reported a preliminary study of pin cherry germination. Apparently, the endocarp or stony shell surrounding the pin cherry seed inhibits germination to some extent, because its removal from recently collected, stratified seeds, combined with appropriate temperature and moisture conditions, results in an average germination of about 45%. However, since the majority of seeds do not respond to this treatment, an additional control mechanism must be present in the seed itself. Marks (1974) proposes a two-step mechanism by which dormancy is ended and germination occurs. The first step involves aging of the endocarp, which breaks down any inhibitors and/or makes it permeable to water and oxygen. The second step involves an environmental sensing mechanism that responds to changes in microclimate, soil or water chemistry, or other factors. This response is even more complicated, for we have found that in the experimentally devegetated ecosystem buried pin cherry seeds continued to germinate for at least four growing seasons. This indicates considerable variation in the capacity of the seed population to respond to altered conditions. Wendel (1972) reported a similar germination pattern for black cherry (*Prunus serotina*) seeds. This response enhances the survival of the species, in that it does not permit the entire seed population to germinate at one time in an all-or-none effort.

A forest fertilization experiment in eastern Pennsylvania suggests that changes in soil chemistry and/or removal of allelopathic substances in cutover areas may trigger pin cherry germination. A series of plots were established in a second-growth black cherry–sugar maple–red maple forest and treated with the following levels of fertilizer: (1) no fertilizer, (2) 300-lb of urea N per acre (1-lb/acre $\times$ 1.12 = 1-kg/ha), (3) 300 lb of urea N and 200-lb of $P_2O_5$ per acre, and (4) 300 lb of urea N, 200-lb of $P_2O_5$, and 100-lb of $K_2O$ per acre (L. R. Auchmoody, personal communication). Nitrogen was applied in two equal applications, at the beginning and in the middle of the growing season. All treatments, with the exception of the control, resulted in the death of the established seedlings of black cherry, sugar maple, and red maple during the first growing season. The following spring, an average of about $1.4 \times 10^6$ pin cherry seedlings/ha appeared on the fertilized plots, while none occurred on the control plots. Auchmoody estimated that there were no pin cherry trees in this forest for at least 40 years prior to fertilization; hence, germination probably came from seeds at least 40 years old.

Since the canopy was not disturbed, changes in soil temperature and moisture or throughfall chemistry cannot be implicated as the cause of germination. Auchmoody points out that the response was about the same for all fertilized plots, which suggests that increased nitrogen in the soil was the key factor, since it was the element common to all treatments. Various workers (Steinbauer and Grigsby, 1957; Anderson, 1968; Roberts, 1969; Popay and Roberts, 1970) have proposed that nitrate is a major regulator of seed germination. However, the death of young seedlings after fertilization indicates that the initial application of fertilizer may have reduced the number of functional roots in the upper soil levels, perhaps temporarily removing allelopathic inhibition of pin cherry germination. It is interesting to note that clear-cutting results in both a markedly elevated concentration of $NO_3$-N in the soil solution and a marked diminution of activity by rootlets of the trees that previously occupied the site. Elevated levels of dissolved substances, such as nitrate or various cations, or elimination of allelopathic root exudates inhibiting seed germination could conceivably play the role of "ectocrines," or "external diffusion hormones," which exert a correlative action on ecosystem function via the external medium (Odum, 1971).

An interesting variation on the buried-seed strategy relates to the stature of species. Examination of openings created by treefall frequently reveals raspberry plants in fairly small openings, but pin cherry seedlings are mostly limited to larger openings. Both species occur abundantly in larger openings, including clear-cuts. The extent to which pin cherry germination commonly occurs in small openings is not well known, but frequent examination of such sites has revealed few pin cherry seedlings, while raspberry seedlings are reasonably common.

Whether or not there exists some discriminating mechanism in these two species which relates germination to the size of the hole in the canopy, the response of each species seems to favor its own survival. Raspberry is a shade-intolerant perennial shrub of small size, usually <1.5 m in height, and its minimum fruiting age is 2 years. Because of its restricted size and early sexual reproduction, raspberry seems to be well adapted to successful sexual reproduction under small canopy openings, where favorable light conditions may exist for only a few years before the opening is closed by outgrowth of surrounding trees and upgrowth of advance-regeneration adapted to intermediate light conditions. Pin cherry, on the other hand, is a shade-intolerant perennial tree that usually grows to a height of several meters and requires about 4 years of growth before fruiting. Rarely are light conditions in small openings sufficient for this type of growth.

## FLORISTIC RESPONSE TO REMOVAL OF THE FOREST CANOPY BY CLEAR-CUTTING

After clear-cutting, most of the strategies described in Table 4-1 become active. The species range from tolerants through intermediates to intolerants (those species that do well only under fairly open conditions). The relative importance of different strategies or species will vary with individual sites, since the response of an ecosystem to clear-cutting is conditioned by a large number of factors, such as time since last cutting, species composition and structure of the ecosystem prior to cutting, nature of cutting, nearness of seed sources, occurrence of good seed years, availability of suitable seedbeds, and microclimate during the recovery period.

To gain information on the early revegetation process, we studied three clear-cut sites during the second growing season after cutting. Sixty 1-$m^2$ plots were distributed equally in the three areas, and all species recorded were assigned to a shade-tolerance class (Table 4-5). We found that revegetation involved a broad spectrum of species, but that localized plots might be dominated by one or two species. There was an average of 28 species per site, and a total of 42 species were recorded. This number would have been higher if species off the plots had been recorded. In another study we laid out 120 1-$m^2$ plots in the center of a 25-m-wide clear-cut strip, and in a stratified-random fashion we harvested one-fifth of these plots before cutting and 1, 2, 3, and 5 years after cutting (S. Bicknell, F. H. Bormann, and P. Marks, unpublished data). Harvesting revealed 20, 29, 26, 26, and 33 species at the respective time intervals, and for the whole period a total of 42 species was recorded.

Species richness, or the number of species per square meter, in the 60-yr-old forest prior to the cut was $5.0 \pm 0.3$ (mean $\pm$ SEM). In a more extensive study of a 55-yr-old forest, Siccama et al. (1970) reported a species richness of $5.5 \pm 0.3$. Species richness rose about 60% the first year

**Table 4-5.** Tolerance Classes and Percentage of Total Cover of Species from Three Northern Hardwood Stands During the Second Growing Season after Clear-Cutting[a]

| Life-form | Tolerance Class[b] | Species | Percentage of Total Cover |
|---|---|---|---|
| Trees | Intolerant | *Betula papyrifera* (white birch) | x[c] |
| | | *Populus tremuloides* (trembling aspen) | 1 |
| | | *Prunus pensylvanica* (pin cherry) | 29 |
| | Intermediate | *Acer pensylvanicum* (striped maple) | 10 |
| | | *A. rubrum* (red maple) | x |
| | | *A. spicatum* (mountain maple) | 1 |
| | | *Betula alleghaniensis* (yellow birch) | 6 |
| | | *Fraxinus americana* (ash) | x |
| | Tolerant | *Abies balsamea* (balsam fir) | x |
| | | *Acer saccharum* (sugar maple) | 7 |
| | | *Fagus grandifolia* (beech) | 6 |
| Shrubs | Intolerant | *Rosa* sp. | x |
| | | *Rubus idaeus* (raspberry) | 16 |
| | | *Rubus* sp. | x |
| | Intermediate | *Sambucus* spp. (elderberry) | 2 |
| | Tolerant | *Lonicera canadensis* | x |
| | | *Viburnum alnifolium* | 3 |
| Herbs | Intolerant | *Anaphalis margaritacea* | x |
| | | *Hieracium* sp. | x |
| | | *Lactuca* sp. | x |
| | Intermediate | *Aster acuminatus* | 2 |
| | | *Dennstaedtia punctilobula* | x |
| | | *Prenanthes* sp. | x |
| | Tolerant | *Arisaema* sp. | x |
| | | *Clintonia borealis* | 1 |
| | | *Dryopteris spinulosa* | 5 |
| | | *Lycopodium complanatum* | x |
| | | *L. lucidulum* | 1 |
| | | *Maianthemum canadense* | x |
| | | *Medeola virginiana* | x |
| | | *Oxalis montana* | 1 |
| | | *Streptopus roseus* | x |
| | | *Trientalis borealis* | 1 |
| | | *Trillium erectum* | x |
| | | *Uvularia sessilifolia* | x |
| | | *Viola incognita* | 1 |
| | | *V. rotundifolia* | x |
| | Unknown | *Carex* sp. | x |
| | | 4 spp. | x |

[a] *Data are based on twenty 1-$m^2$ plots systematically distributed in each stand. All sampled vegetation was <2.5-m tall (P. Murphy, personal communication).*
[b] *See Table 4-2. Tolerance classes for shrubs and herbs are based on Siccama et al. (1970) and personal observations.*
[c] *x = <0.5%.*

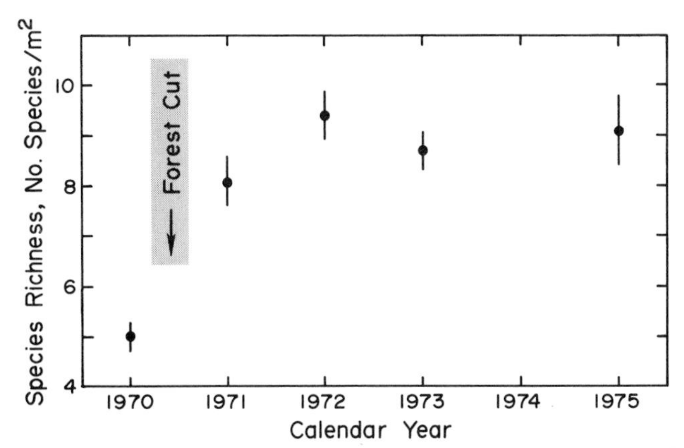


**Figure 4-3.** Species richness based on individuals rooted in 1-m² plots in a 60-yr-old northern hardwood forest in 1970 and in subsequent years after clear-cutting. Means for each year are based on 19 to 22 randomly distributed plots. The range shown is the standard error of the mean (S. Bicknell, F. H. Bormann, and P. Marks, unpublished data).

after the cut and leveled off to a slightly higher value by Year 2 (Figure 4-3).

In both studies the flora was a thorough mixture of tolerance classes. Tolerant and intermediate species relatively abundant in the uncut forest were most common but were thoroughly mixed with intolerant species (buried-seed species and adventive species) that had been unimportant previously (Table 4-5).

## DIFFERENTIATION OF THE VEGETATION ESTABLISHED IMMEDIATELY AFTER CLEAR-CUTTING

The tangled vegetation that follows within a year or two of clear-cutting, generally <2-m high but rich in species and in numbers of individuals, may be considered as a single undifferentiated stratum. The process of differentiation begins very rapidly. Some individuals grow faster, some can survive in the shade of others, and an overstory, understory, and herb layer develop fairly rapidly.

During the first 1 or 2 years after clear-cutting, dominance is often shared by woodland herbaceous species (e.g., *Aster acuminatus*, *Dennstaedtia punctilobula*, and *Uvularia sessilifolia*) whose clones may expand rapidly, herbaceous adventives such as *Epilobium*, and a tangle of woody vegetation composed of sprouts, shrubs, and seedlings and saplings of trees from all tolerance classes (Wilson and Jensen, 1954; Marquis, 1965; Horsley and Abbott, 1970). The trees that grow up, after clear-cutting can be considered even-aged (Gilbert and Jensen, 1958; Marquis, 1967), since the new seedlings and sprouts and the release of established individuals

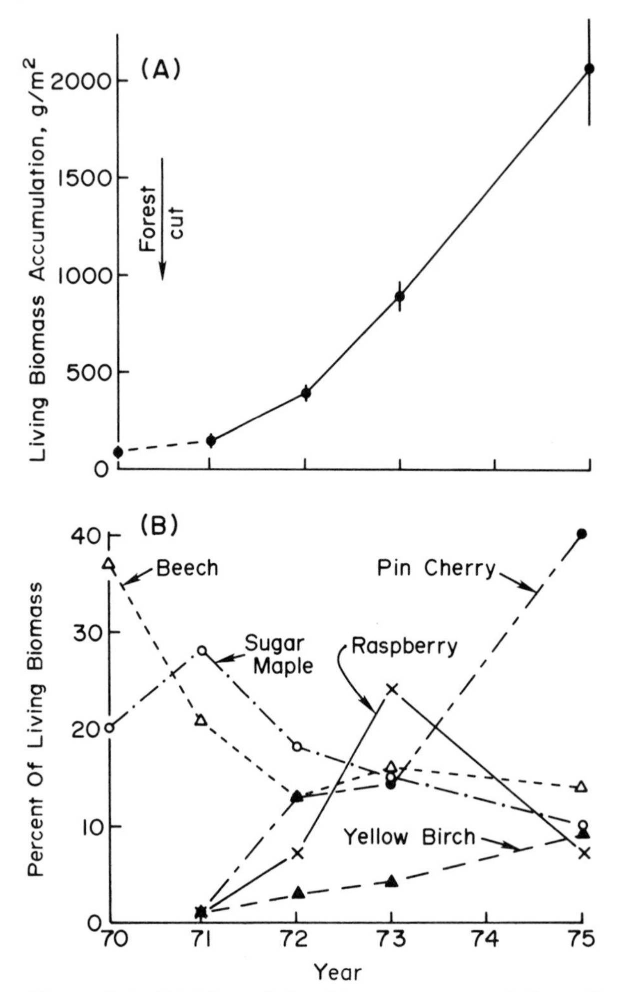


**Figure 4-4.** (A) Mean living biomass accumulation after clear-cutting in grams per square meter (range is the standard error of the mean); data for 1970 represent biomass in the herbaceous layer of the uncut forest, while subsequent data represent all biomass on the 1-m² plots. (B) Percentage of living biomass found in various species at different times after clear-cutting; not all species are shown (S. Bicknell, F. H. Bormann, and P. Marks, unpublished data).

date to within a few years of each other. This is a fact of paramount importance, since clear-cutting imposes a more or less even-aged condition on the site, and within a few years the site is occupied by populations of ıdividual plants that have the potential of dominating the area for more than a century without recruitment of new seedlings.

Differentiation during the first 5 years after clear-cutting is relatively rapid, and dominance, based on biomass accumulation, may pass rapidly from one group of species to another. This pattern was seen clearly in the plots harvested before cutting and at intervals afterward (see above).

After cutting, the density of individuals rose rapidly, peaked in about the second year, and then declined. Accumulation of living biomass rose exponentially until sampling was terminated in the fifth year (Figure 4-4A). However, the proportions of living biomass in various species shifted rapidly (Figure 4-4B). Prior to the cut, advance regeneration of beech and sugar maple dominated the layer. During the first and second year after cutting, advance regeneration of sugar maple was the major component of the living biomass, the third year raspberry was predominant, and by the fifth year pin cherry was clearly the most important species.

Relative height growth is one of the major differentiating factors, and differences among tolerance classes appear quickly. This factor was already evident in the 2-yr-old stands discussed above. Fast-growing shade-intolerant tree species, although having a density similar to that of the tolerant and intermediate classes, comprised the majority of the tallest individuals (Table 4-6).

Differentiation of northern hardwood stands subsequent to disturbance has been intensively studied by forest scientists in the Hubbard Brook region (e.g., Jensen, 1943; Wilson and Jensen, 1954; Marquis, 1965), and the following pattern emerges. Often, within a few years after clear-cutting, dominance shifts to fast-growing tree species. These include shade-intolerant species, such as pin cherry, trembling aspen, bigtooth aspen, and white birch, and species of intermediate tolerance, such as ash, red maple, and striped maple. These individuals tend to grow taller than the general mass of the vegetation, and this results in stratification of the forest canopy into an overstory and understory. Species such as sugar maple, beech, red spruce, and yellow birch are frequently represented in large numbers but occur mostly as codominants or lower-crown classes. Continued differentiation is in many respects an age-dependent phenomenon, with dominance progressively shifting from shorter-lived, faster-growing, less-tolerant species to longer-lived, initially slower-growing, more-tolerant species (Jensen, 1943; Marquis, 1967).

**Table 4-6.** Relationship of Tolerance Class, Density, and Stem Height for Trees in a 2-Yr-Old Northern Hardwood Forest[a]

| | Density (No. of Stems/m$^2$) | | Stem Height | |
|---|---|---|---|---|
| Tolerance Class | Seedlings | Sprouts | <50-cm (%) | >50-cm (%) |
| Intolerant | 15.2 | 0.3 | 38 | 59 |
| Intermediate | 12.6 | 0.7 | 39 | 26 |
| Tolerant | 5.5 | 2.4 | 23 | 15 |

[a] *Data based on twenty 1-m$^2$ plots in three sites (P. Murphy, personal communication).*

Pin cherry will die about 30 years, and aspen between 40 and 60 years after cutting. The death and fall of pin cherry and aspen or smaller individuals of white birch, red maple, and ash create relatively little disturbance at the level of the forest floor, and the space created in the canopy is usually filled by encroachment of dominants and codominants from lower-crown classes of intermediate and tolerant species. These individuals are of approximately the same age as those they replace. The process of canopy closure by surrounding trees is well known for even-aged temperate forests in Europe (Jones, 1945). White birch may remain for about 80 years after clear-cutting, and red maple and ash may survive more than 100 years before they die out (Marquis, 1967). Many of these trees may be replaced by lower-crown classes of shade-tolerant trees of about the same age. Some trees, however, because of their larger size, may create larger openings when they fall. These openings may be filled by seedlings or saplings from advance regeneration on the forest floor that became established well after clear-cutting. The production of gaps and their closure by growth from the forest floor represents the beginning of a process that will become more prevalent with time and by which the stand ultimately will become truly uneven-aged.

Given no exogenous disturbance, after about 60 years of the Aggradation Phase we would expect the ecosystem to exhibit these characteristics: (1) The overstory and understory are more or less continuous, i.e., there are relatively few gaps, and are dominated largely by even-aged individuals dating from the time of cutting. (2) As a result of the natural thinning process, the over- and understories have become increasingly dominated by tolerant and intermediate species that have the potential to remain alive for long periods. Shorter-lived intolerant species essentially have been eliminated as active growing components, and intermediate species have become reduced in importance. (3) Openings in the forest caused by the death and fall of individual trees are generally of small to modest size (e.g., 50 to 150-$m^2$) and are filled primarily by upgrowth from the forest floor of advance regeneration of tolerant, along with some intermediate species. This process, discussed by Forcier (1973), contributes to a further increase in the importance of tolerant species. (4) The density of dominant trees declines, but their diameter at breast height, crown diameter, and biomass increase as the survivors become more massive. The subsequent fall of these larger trees tends to create larger openings that favor higher proportions of intermediate and shade-intolerant species in the new canopy which fills the opening.

## GROWTH STRATEGIES UNDERLYING INITIAL CANOPY DIFFERENTIATION

Before resuming consideration of the phytosociological development of the ecosystem, we shall pause to examine some of the growth strategies that allow species to occupy the different stages of ecosystem development that follow clear-cutting. If no exogenous disturbance occurs, the shifts in dominance among tolerance classes after clear-cutting often follow the generalized pattern shown in Figure 4-5 (Marks, 1974; Aber, 1976).

The rewards of dominance are great, since dominance implies a proportionately larger control over biogeochemical and hydrologic pathways of the ecosystem, a larger share of available nutrients and water, and, through canopy position, a major share of the radiant energy of the site. Why should any group surrender dominance once it is achieved? The answer resides in the fact that continued growth is itself constantly altering the conditions of selection that exist within the ecosystem, and no one higher plant species or group of species has solved the problem of coping with the entire selection spectrum.

Although there are infinite variations on the life-strategy theme, it is useful to consider the generalized curves of Figure 4-5 as reflecting two opposing strategies: an exploitive strategy best adapted to conditions immediately after clear-cutting and a conservative strategy better adapted to conditions in the mid-Aggradation Phase, about 70 to 100 years after clear-cutting. No one species is wholly exploitive or conservative. All share, to varying degrees, aspects of both strategies. Among tree species, pin cherry may be considered as most closely approaching a purely exploitive strategy, whereas beech might be considered a thoroughly conservative species. Similar patterns of plant adaptation or species strategies have been implicated or proposed in a number of other studies (Grime, 1974; Harper and White, 1974). Harper and White's (1974) concept of shade-intolerant, colonizing (r-type) species, for instance, is closely related to the exploitive growth strategy. By way of contrast, shade-tolerant species possessing fewer and larger seeds and a long juvenile period (Harper and White, 1974) could be considered representative of the conservative strategy.

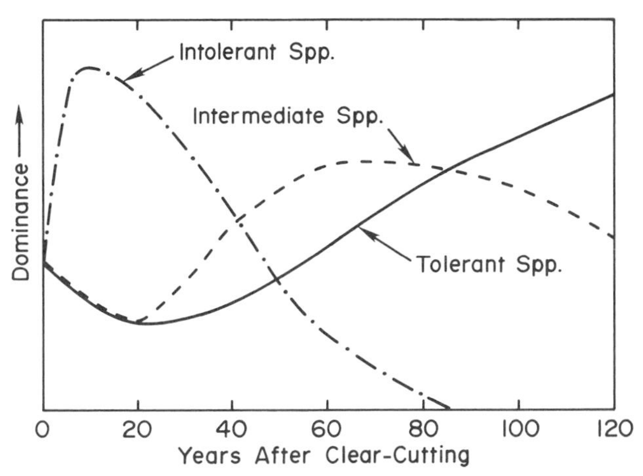


**Figure 4-5.** A hypothetical sequence showing trends in dominance among different shade-tolerance classes of trees after clear-cutting.

In relation to ecosystem development, the exploitive strategy may be thought of as "getting while the getting is good." By a combination of reproductive, developmental, and physiological growth attributes, the exploitive species are often able to attain local dominance rapidly on the cutover site, which is temporarily enriched with available nutrients, water, and radiant energy. This idea will be developed more fully in subsequent sections of Chapter 4 and in Chapter 5. Rapid attainment of dominance is often coupled with rapid sexual reproduction and a next generation of progeny able to exploit other temporarily enriched situations as they arise.

The exploitive strategy carries costs as well as benefits. Costs involve the relative inability to compete successfully under conditions of stress, for example, in shade or where there is strong competition for available water and nutrients. Often, species favoring the exploitive strategy are relatively short-lived, and, even though they attain temporary dominance with all of the rewards that entails, they ultimately surrender the site to longer-lived more-competitive species.

Cutover ecosystems with high rates of NPP (Chapter 5) dominated by species such as pin cherry and raspberry tend to be characterized by accelerated losses of water and nutrients. This, coupled with poorly developed root systems (low root/shoot ratios, Table 4-7), suggests that species employing the exploitive strategy may be less efficient in the use of water and nutrients than those using the conservative strategy.

The conservative strategy proposed here is one in which physiological and developmental aspects of growth are not geared to a perturbed and rapidly changing situation but rather to a relatively constant and predictable environment, such as the one that typifies the mid-aggrading forest discussed in Chapter 2. In this strategy we might think that nutrient uptake is closely coupled to the fairly constant patterns of decomposition and mineralization that presumably exist in the aggrading forest, particularly after the forest floor has achieved equilibrium (Figure 2-1). Conservative species would be strongly competitive at the low end of the range of availability of water and nutrients and would tend to use these resources fully. The conservative strategy also involves the capacity to use radiant energy efficiently, particularly at low intensities. Dominance of the ecosystem by species employing the conservative strategy might contribute to the low nutrient and stream-water export that typifies mid-aggrading ecosystems.

The conservative strategy also has its costs. Apparently, a strategy that couples a species' growth to fairly constant and predictable mineralization and a wide range of light conditions limits its capacity to respond quickly to the conditions of site enrichment that follow clear-cutting. This limitation may be most apparent in seedlings of tolerant species genetically equipped to survive the highly competitive conditions of the forest floor under a closed canopy. Several workers (Grime, 1965; Parsons, 1968) have proposed the idea that selection for slow growth

**Table 4-7.** Morphogenetic Characteristics of Saplings of Northern Hardwood Species[a]

| | Extension Growth of Leader | | | | |
|---|---|---|---|---|---|
| Species[b] | Type | Days to >90% growth | Length (cm) | Heterophylly | Root/shoot ratio[c] |
| Intolerant species[d] | | | | | |
| White birch | Indeterminate | 112 | — | Yes | — |
| Gray birch | Indeterminate | 60 | — | — | — |
| Bigtooth aspen | [Indeterminate][e] | 60 | 70–100 | Yes | — |
| Trembling aspen | Indeterminate | 60 | >100 | Yes | $0.16 \pm .02$ |
| Pin cherry | Indeterminate | 100 | >100 | Yes | $0.19 \pm .05$ |
| Intermediate species | | | | | |
| Red maple | Indeterminate | 40 | 40–50 | Yes | — |
| Ash | Determinate | 30 | >60 | No | — |
| Yellow birch | Indeterminate | 60 | >43 | Yes | — |
| Tolerant species | | | | | |
| Sugar maple | [Determinate] | 30 | 25–40 | No | $0.37 \pm .16$ |
| Beech | Determinate | 30 | 25–40 | No | $0.36 \pm .09$ |
| Hemlock | | 85 | 45 | — | — |

[a] *Adapted from Marks (1975).*
[b] *Arranged according to tolerance classes.*
[c] *Values are means ± SEM.*
[d] *These tolerance classes are essentially the same as Marks' successional classification, i.e., intolerant equals early successional and tolerant equals late successional. The one exception is yellow birch, which Marks classifies as late successional.*
[e] *Brackets indicate suspected status.*

rates is a frequent adaptation to limiting levels of resource availability. Seedlings of eastern hemlock, perhaps the most shade-tolerant of northern hardwood species, have been cited as having low metabolic and growth rates even under favorable experimental conditions (Grime, 1965).

In clear-cuts, species programmed for the conservative growth strategy fall behind those using the more responsive exploitive strategy and soon suffer the problems associated with a subdominant position. However, because of their tolerance of stressful situations and their potential to live a long time, individual trees can survive and eventually dominate. Their reproductive strategies are such that they can selectively reinforce their populations in the overstory when modest openings are created by the death of other trees. By virtue of these attributes, species employing a conservative growth strategy eventually come to dominate the cutover site, barring exogenous disturbance.

After clear-cutting of a northern hardwood ecosystem, species employing the exploitive strategy are able to respond more rapidly than

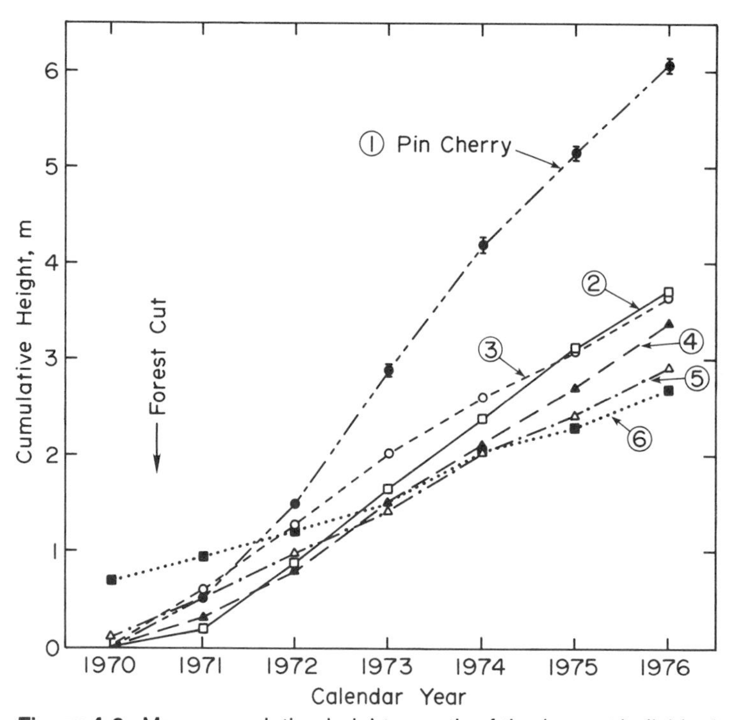


**Figure 4-6.** Mean cumulative height growth of the largest individual of each tree species found on each of fifty 30-m$^2$ plots in a 6-yr-old clear-cut. The 60-yr-old forest was cut in the fall of 1970. Height in 1970 represents advance regeneration before cutting. (1) Pin cherry; (2) trembling aspen; (3) striped maple; (4) yellow birch; (5) sugar maple (mountain maple and ash followed about the same pattern); (6) beech. Data are the means of 15 to 50 individuals. The range shown for pin cherry is the standard error of the mean.

conservative species occupying the same site. An example of this is shown in the 6 years of growth after a clear-cut at Hubbard Brook (Figure 4-6). Height growth of pin cherry has clearly outstripped that of other species on the same plots. In general, height growth responses are related to tolerance class, with intolerant species highest, intermediate species in the middle, and tolerant species lowest. It is interesting to note that beech and sugar maple, among the most conservative species, were the least responsive, even though both had average heights equal to or greater than most other species at the end of the first growing season after cutting.

John Aber, in a recent (1976) unpublished review of the characteristics of early- and late-successional northern hardwood species, has pointed out that not only do early (exploitive) species have faster growth rates but their growth rate is age related, with highest rates early in an individual's lifespan. This is shown by the annual rate of height growth for pin cherry in the 6-yr-old cutover site (Figure 4-7). Pin cherry in the third and fourth years shows growth rates averaging about 1.3-m/yr, but this falls to about 0.95-m/yr in Years 5 and 6. One interpretation is that the early burst of height growth allows the species to gain local dominance quickly, partially shade its competitors, and get a proportionately higher share of local site resources. Thereafter, rapid height growth is of less strategic value, since the species has temporarily gained a better competitive position and other species must live under conditions it creates. The leveling off after 1973 of annual growth rates of the other species might reflect the somewhat inferior position of these species within the canopy. In an analysis of annual growth rates over a 90-yr span, Aber has shown that white birch (early successional) achieved 79% of its total growth before Year 20, while sugar maple achieved only 47% of its growth during the same period.

Our data suggest that the soil solution in the clear-cut site is enriched with nutrients shortly after cutting, and the question arises as to the responsiveness of species to this temporary condition. No direct answer is possible, but indirect evidence suggests that at least three species are capable of responding to this condition with greatly accelerated growth rates.

Safford and Filip (1974) fertilized (4500-kg of limestone and 1200-kg of commercial 15-10-10 NPK fertilizer per hectare) plots after clear-cutting of a northern hardwood forest and 4 years later found an increase in standing crop of live biomass from 7-t/ha on control plots to 24-t/ha on fertilized plots. Leaf biomass increased from 980 to 1780-kg/ha. Almost all of this increase was due to vigorous pin cherry growth. Pin cherry biomass increased from 41% of the total on control plots to 89% on fertilized plots. These results suggest that at least part of pin cherry behavior on cutover sites is exploitation of the temporary nutrient enrichment discussed in Chapter 3.

Tamm (1974) demonstrated a similar result with raspberry (***Rubus***

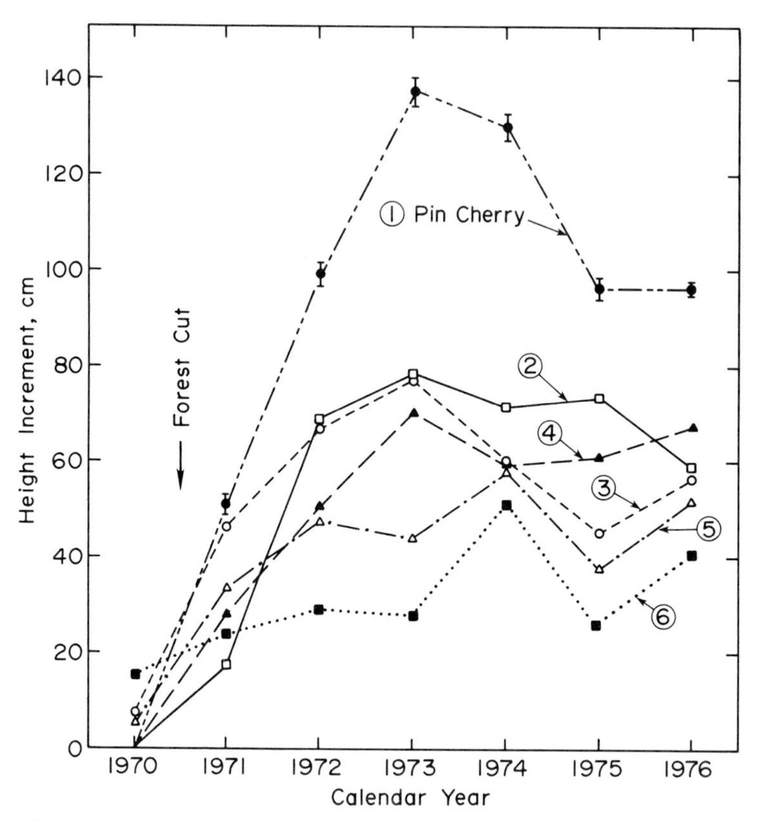


**Figure 4-7.** Mean annual height increment of the largest individuals on fifty 30-m² plots in a 6-yr-old clear-cut. The 60-yr-old forest was cut in the fall of 1970, and height in 1970 does not represent mean annual growth but rather the sum of the height due to advance regeneration divided by the total number of individuals found on all plots. (1) Pin cherry; (2) trembling aspen; (3) striped maple; (4) yellow birch; (5) sugar maple; (6) beech. Data represents the means of 15 to 50 individuals. The range shown for pin cherry is the standard error of the mean.

*idaeus*) in Sweden. Combinations of nitrogen fertilizer ($NH_4NO_3$), ranging from 0 to 180-kg N/ha-yr, and phosphorus (Superphosphate 26/30), ranging from 0 to 40-kg P/ha-yr, were applied in 24 treatments to plots located in a spruce plantation established on podzol soil in 1957. Fertilizers were applied over a series of years between 1967 and 1973. Aboveground dry-weight production of raspberry in 1973 was up to two orders of magnitude greater on the more heavily fertilized plots. On control plots, raspberry was absent or a minor vegetational element constituting about 1% of the annual aboveground dry-weight production. In contrast, raspberry constituted up to 8% of the annual production on fertilized plots, even though fertilization promoted stand closure, which probably diminished the amount of light reaching the forest floor. Increases in raspberry were associated with a 5- to 20-fold increase in available nitrogen in the soil, including some $NO_3$-N, which was virtually absent on the control plots.

Ash is usually considered an intermediate species with regard to light tolerance, but on 3- to 5-yr-old clear-cuts we have observed instances of unusual height growth sometimes equaling that of pin cherry. In this regard, it is interesting to note that Mitchell and Chandler (1939), in a pioneering study of nitrogen nutrition among hardwood species, found white ash to be among the most nitrogen-demanding species. Based on a reanalysis of Mitchell and Chandler's data, Aber suggests that the growth of ash is very responsive to nitrogen availability. This behavioral pattern suggests that the very rapid early growth observed for ash on some cutover sites may be an exploitive response to the rapid increase in nitrogen availability (nitrate) after cutting.

## MORPHOGENESIS AND GROWTH STRATEGY

Marks (1975) has developed an important hypothesis linking morphogenesis and successional behavior of northeastern trees. Basically, he has shown that early-successional trees have a growth regime well adapted to the exploitive strategy while late-successional species do not.

Even in well-lighted environments, shade-tolerant late-successional species achieve relatively modest amounts of height growth and generally complete height growth early in the growing season (Table 4-7). Early-successional species achieve more height growth by extending their period of growth over longer periods of time, often ceasing growth only 2 to 4 weeks before the first autumn frosts.

Shoot morphogenesis is one of the bases for these contrasting patterns of growth. There are two general ways in which terminal shoots are formed in temperate woody angiosperms. In *determinate* growth the shoot is fully preformed in the winter, with embryonic leaves and leaf primordia, and extension growth during the following season is usually limited to a fixed number of leaves determined when the bud was formed in the previous growing season. In *indeterminate* growth, the winter bud carries a few leaves, and these expand as the bud opens and extension growth begins in the spring. The shoot apex, however, remains active and continues to initiate new leaves and internodes during the growing season.

Apparently the indeterminate growth pattern is closely associated with heterophylly, in which leaves formed in the winter bud are morphologically distinct from those formed during the current growing season (Critchfield, 1960). Extension or height growth is thus fairly well fixed during the previous growing season in species with determinate growth, while species with indeterminate growth have the capacity to produce new leaves and internodes as long as the environment during the current growing season is favorable.

Marks' data indicate that early-successional, or shade-intolerant, trees in the northeastern region utilize the indeterminate growth pattern, while shade-tolerant late-successional trees generally exhibit the determinate pattern (Table 4-6).

Two other aspects of shoot morphology not considered by Marks relate to branching and the apical dominance of the terminal bud over axillary buds formed the same year. Whitney (1976) reports that twigs of late-successional species (such as beech and sugar maple) tend to bifurcate, producing a dorsiventrally flattened, two-ranked branching pattern that minimizes leaf overlap more than that of early-successional species. In contrast, early-successional species (such as aspen) bifurcate less, producing erect long shoots bearing numerous short branches. In effect, leaves are displayed in a columnar volume characterized by much overlap but considerable side lighting. In contrast to Whitney's findings, we have observed instances in which the terminal bud of pin cherry did not inhibit the extension of lateral buds formed during the same year—the generally recognized case. This phenomenon, which may be a response to especially enriched situations, leads to secondary and even tertiary branch formation during the same year. This may contribute to the high leaf-area indices reported by Marks (1974) for young pin cherry stands.

Marks (1975) reported preliminary data on root/shoot ratios of early- and late-successional species. Ratios for saplings of trembling aspen and pin cherry were substantially below those for saplings of sugar maple and beech (Table 4-7). Marks noted that root systems of pin cherry and aspen growing on cutover sites were shallow (generally less than 35-cm deep), much branched (often with major lateral roots within 2 to 4-cm of the surface), and easily pulled from the soil by a single person (trees as tall as 7-m!). He suggests that this rooting habit is especially geared toward temporary occupancy of the site, with energy preferentially invested in aboveground structure while roots are relatively weak and concentrated in the forest floor.

Finally, an interesting aspect of the indeterminate growth pattern relates to nutrient resorbtion from leaves just before leaf fall (Ryan, 1978). In determinate species, concentrations of nitrogen and phosphorus within the leaves begin to decrease about a month prior to leaf drop. These nutrients are withdrawn into perennial portions of the tree and stored during the dormant season. During the next growing season resorbed nutrients may account for as much as 40% of the total annual growth requirements for both nitrogen and phosphorus. In the indeterminate species pin cherry a somewhat different pattern of nutrient resorption prevails. Although a general resorption of nutrients occurs prior to leaf drop, the phenomenon apparently goes on throughout much of the growing season. As extension growth occurs and new leaves are produced by the meristem, some of the leaves formed earlier senesce. A rough estimate suggests that about 10% of all leaves formed senesce prior to the maximum period of senescence. Apparently, nutrients are withdrawn and transported to growing points. Thus, resorption may result in a proportion of the resorbed nutrients being used at least twice in the same growing season in the support of new growth.

Indeterminate growth lies at the base of the exploitive strategy. Shoot growth and leaf formation extend well into the growing season and depend to a considerable degree on current photosynthesis; specifically, photosynthate from the first-formed leaves supports the continued mitotic activity of the shoot apex and the expansion of new shoots and leaves during the growing season (Kozlowski and Keller, 1966; Marks, 1975). Thus, extension growth in indeterminate species is largely determined by conditions of the current growing season. This allows the species to capitalize on any temporary enrichment of the site that may exist. But indeterminate extension growth is not only a means of capitalizing on temporary site abundance; it also may be an effective mechanism for directing crown development toward areas where light is most abundant (J. Aber, unpublished). The leaves of intolerant species are thought to be light saturated only at higher light intensities; hence, twigs in higher-light environments would tend to extend more. Indeterminate growth fueled by favorable conditions of light, water, and nutrients can lead to the rapid height growth and early dominance discussed previously.

Determinate growth would seem to be a basic component of the conservative growth strategy. Not only are the numbers of leaves determined in the previous growing season, but most extension growth draws on carbohydrates stored in previous seasons (Kozlowski and Keller, 1966; Marks, 1975). This pattern of growth has advantages in highly competitive situations that are apparent. Extension growth and leaf expansion occur very rapidly in the early growing season, largely using stored energy. This permits species to exploit the temporarily favorable environment at the beginning of the growing season. Generally speaking, available water supplies have been fully recharged during the spring snowmelt period, and, owing to continued decomposition and mineralization during late autumn, winter, and the early spring months, seasonal nutrient availability also may be at a maximum (Muller and Bormann, 1976). Melillo (1977) has followed the seasonal course of mineralization using forest-floor soil samples placed in polyethylene bags and returned to the soil profile. His data show maximum mineralization in the early part of the growing season, i.e., June. The determinate growth pattern allows species to draw quickly on this seasonal enrichment. Thereafter, productivity in closed forests is largely a function of the seasonal weather pattern, which determines light, temperature, and soil moisture regimes and probably influences rates of decomposition. The determinate growth pattern would seem to be well adapted to situations of intense competition, in which the range of variables controlling growth is relatively modest (at least in comparison to disturbed situations) and a premium is put on rapid early growth. Apparently, most photosynthate is used to support radial and root growth or to refurbish supplies of stored energy rather than being invested in extension growth. The determinate growth pattern, coupled with a photosynthetic apparatus able to use a wide range of light intensities and a relatively larger investment of energy in root growth (Table 4-7), would seem to be well adapted to closed forest situations typified by intense competition for light, water, and nutrients.

Aspects of the exploitive and conservative growth strategies discussed above are summarized in Table 4-8.

**Table 4-8.** Aspects of Exploitive and Conservative Growth Strategies as Found Among Tree Species of the Northern Hardwood Ecosystem[a]

| | Growth Strategy | |
|---|---|---|
| Aspect | Exploitive | Conservative |
| Longevity | Short | Long |
| Extension Growth | | |
| Morphologic type | Indeterminate | Determinate |
| Time of occurrence during growing season | Throughout | Early |
| Height Growth | | |
| Annual rate | Fast | Slower |
| Period of greatest growth | Early years | Later |
| Heterophylly | Yes | No |
| Bifurcation ratio | High | Low |
| Crown Characteristics | | |
| Width | Narrower | Wider |
| Depth | Shallower | Deeper |
| Root Growth | | |
| Root/shoot ratio | Lower | Higher |
| Depth of rooting | Shallower | Deeper |
| Photosynthetic light-saturation of individual leaves | At higher intensities | At lower intensities |
| Absorbtion of water and nutrients at lower range of availability | Less effective | More effective |
| Reproductive strategy | | |
| Time to first fruiting | Shorter | Longer |
| Advance regeneration | No | Yes |
| Seed weight | Very light to heavy | Relatively heavy |
| Buried-seed strategy | Yes | Not well developed |
| Fugitive-seed strategy | Yes | No |

[a]*Many items are expressed in relative terms. After Marks, 1975; Aber, 1976, and unpublished data; Whitney, 1976; and personal observations.*

## ENDOGENOUS DISTURBANCE

In our discussion of ecosystem development we have deliberately omitted any mention of the effects of exogenous disturbances like cutting, hurricanes, fire, or ice storms. These stochastic events are not uncommon and, in fact, are major determinants of landscape pattern. However, our goal in the first six chapters is to theorize on how animate and inanimate components interact to produce ecosystem development in the absence of outside (exogenous) disturbances.

Up to this point we have made little mention of endogenous disturbance, but there is at least one source of disturbance that is inherent in the forest ecosystem and which cannot be ignored or even theoretically set aside. That force is treefall. When a tree falls, it creates a local disturbance the size and intensity of which is related to the size of the individual. The potential to create disturbance in this way varies markedly among species since only those species capable of achieving relatively large size and weight are capable of creating a large local disturbance. Many shade-intolerant species such as pin cherry, trembling aspen, bigtooth aspen, gray birch, and white birch are genetically incapable of achieving the size necessary to create large openings when they fall. On the other hand, a number of intermediate and shade-tolerant species have this capacity (Figure 4-8).

We have computed the average aboveground fresh weight for the three species that often dominate mid- to late-aggrading forests (Figure 4-9). Two trends are evident from these data. As a tree increases in size, its fresh weight increases exponentially, and the proportion of weight localized in the crown seems to increase linearly. A tree in the 75- to 100-cm class weighs from 8 to 12-t (fresh weight), and about 40% of the weight is concentrated in the branches. This represents a large amount of potential (mechanical) energy concentrated at one point in the ecosystem, and the high proportion of weight in the crown would amplify the tree's capacity to do damage when it falls over.

To gain an appreciation of the amount of damage a large tree can cause when it falls, we looked at a number of recently fallen trees in well-developed forests. Three are reported here. In the Beinecke Forest in Williamstown, Massachusetts, Henry Art called to our attention a 140-cm-dbh red oak, about 30 m tall, that had fallen 3 years previously. This tree not only vacated its own space in the canopy, but in the process of falling broke the top out of a 36-cm-dbh sugar maple and uprooted two sugar maples 30 and 15-cm in dbh. This created a hole greater than 300-$m^2$ in the upper canopy and a well-lighted area of about 150-$m^2$ on the forest floor. Seedlings and saplings of all tolerance classes were vigorously growing in the opening. At Smith's Woods in Trumansburg, New York, Peter Marks led us to an opening in a well-developed forest where a 100-cm-dbh beech, about 28-m tall, was uprooted; in the process of falling it knocked over a 10- and 40-cm-dbh sugar maple and a 40-cm-dbh

**Figure 4-8.** Three white ash trees blown over in a wind storm. The fall illustrates the "domino effect"—the first tree crashed into the second and the second into the third. In this way, the fall of a large individual tree can create a canopy opening much larger than the space formerly occupied by its own crown. The fall shown here created a hole in the overstory canopy about 35-m long and 5–15-m wide. Upturned roots were up to 1.8-m aboveground, and pits were made as deep as 1-m.

basswood. An opening of about 300 $m^2$ was made in the upper canopy, and seedlings and saplings of all tolerance levels were found growing in the vacated space on the forest floor. At Campton, New Hampshire, we observed a 75-cm-dbh white pine that had snapped at about 3 m above the ground. As it fell, the pine stripped most of the limbs from a 40-cm-dbh sugar maple, snapped an 18-cm-dbh sugar maple and a 15-cm-dbh beech, and uprooted a 38-cm-dbh beech. The tangle of branches of the fallen trees flattened most of the saplings in a substantial area. It was estimated that this fall opened an area in the upper canopy of approximately 500-$m^2$. This opening was observed early in the first summer after the tree fell over, and vegetative response to the disturbance was not yet clear.

It is clear that treefall is a major and unavoidable source of endogenous disturbance since "what goes up must come down." The amount of disturbance is related to the biomass of the falling individual(s), and openings created in the canopy may range from a few to hundreds of square meters. In its maximum expression individual treefall may involve: (1) creation of a throw mound(s) which produces a new microtopography as well as a variety of new seedbeds (Hutnik, 1952) and (2) substantial

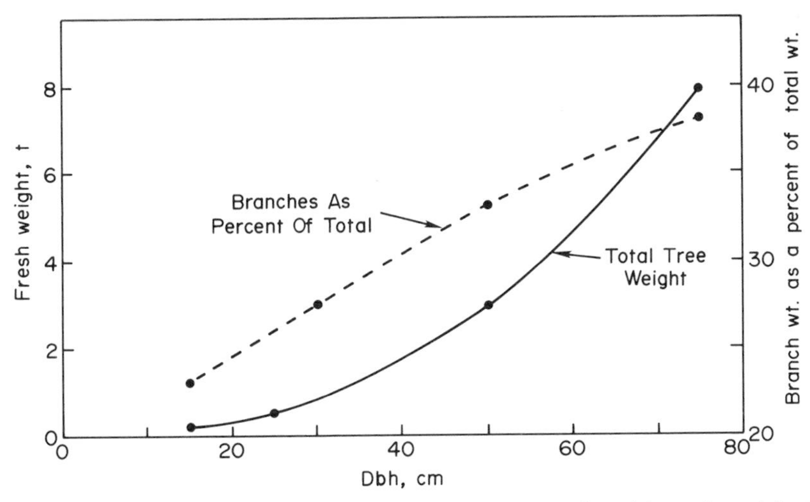


**Figure 4-9.** Estimated total fresh weight aboveground and branch weight (percentage of total weight) as a function of dbh. Data are an average of total weight for yellow birch, sugar maple, and beech. Estimates were made using regression equations for stem, branch, leaf, and twig dry weight (Whittaker et al., 1974) and fresh moisture contents (USDA Publ. No. 72).

openings in the upper canopy, e.g., 300 to 500-$m^2$, comprising space vacated by the falling tree plus that resulting from snapping off, uprooting, or limb-stripping of individuals in its path of fall.

Thus, treefall may have a marked effect on the biogeochemistry of local points within the ecosystem. Treefall implies a local reduction in net primary productivity, an increase in dead wood, and an accumulation in the soil of water and nutrients previously used by the fallen trees. It also implies increased decomposition and mineralization due to increased water and nutrients and warmer soil. In other words, the location becomes an enriched site within the ecosystem. Of course, the degree of enrichment is related to the amount of disturbance caused by the fall, and this is related to tree size. It seems probable that there is a continuum of biogeochemical, hydrologic, and radiant energy responses at the forest floor, ranging from those similar to a closed aggrading forest (Chapter 2) when a small tree falls to those approaching a recently clear-cut forest (Chapter 3), when a patch of large trees fall.

## INTERACTIONS BETWEEN REPRODUCTIVE STRATEGIES AND DEGREE OF CANOPY DISTURBANCE

Based on our own observations and extensive patch and clear-cutting experiments by scientists of the U.S. Forest Service, we suggest that for northern hardwood forests, the importance of the various reproductive strategies outlined in Table 4-1 is differentially related to the size of holes created in the canopy. We express these relationships in a rough way in Figure 4-10, which shows the response of tree species in disturbed areas, ranging from areas a few square meters in size to clear-cuts. In the next section, we shall present a generalized pattern of secondary succession that relates canopy differentiation to the creation of larger and larger holes and to a shift in the relative importance of the various reproductive strategies. These processes culminate in the formation of an all-aged forest, the end point of the developmental process.

Figure 4-10 is intended to illustrate, in a general way, responses of different reproductive and growth strategies to a gradient of canopy disturbances. Four categories of growth by which the dominant canopy may be reestablished are proposed. These are:

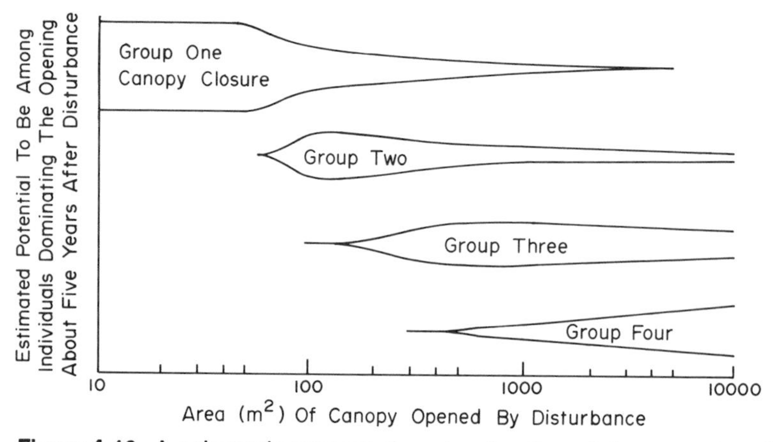


**Figure 4-10.** A schematic presentation showing the relative capacity of various tree species to achieve dominance within about 5 years in newly created disturbed areas of various sizes in second-growth northern hardwood forests. Tolerant and some intermediate individuals that do not gain immediate dominance may remain as lower crown classes and later gain a dominant position when an earlier dominant dies out. Groups 2, 3, and 4 represent advance regeneration, seedlings or saplings, and/or new seedlings that grow upward from the herb–shrub strata, while Group 1 represents canopy closure by lateral or upward growth by surrounding trees. Group 2 is composed of shade-tolerant species and includes beech, sugar maple, hemlock, red spruce, and balsam fir. Group 3, the intermediate species, contains red maple, striped maple, mountain maple, yellow birch, and ash. Group 4, the shade-intolerant species, includes pin cherry, white birch, gray birch, bigtooth aspen, and trembling aspen. The scheme is based on observations of growth in various-sized openings (50 to about 500-$m^2$) caused by the death and fall of various-sized trees and literature reports on forest cuts ranging from 400 to 10,000-$m^2$. Estimates are based on personal observations and data from Jensen, 1943; Hutnik, 1952; Wilson and Jensen, 1954; Gilbert and Jensen, 1958; Leak and Wilson, 1958; Blum and Filip, 1962; Marquis, 1965; Horsley and Abbott, 1970; and Bjorkbom, 1972.

1. *Canopy closure by surrounding trees.* These may be dominant or codominant individuals of shade-tolerant, intermediate, or intolerant trees or lower-canopy classes of tolerant or intermediate species. These individuals fill the gap by growth from the side or from below and may completely fill small openings in the canopy (ca. $50\text{-m}^2$).
2. *Upgrowth by shade-tolerant species.* This category includes growth by new seedlings and advance regeneration up to several decades in age that can respond to openings from about $50\text{-m}^2$ to those the size of clear-cuts and grow upward to fill the space created in the canopy.
3. *Upgrowth by intermediate species.* This includes growth by new seedlings and advance regeneration generally less than ten years old. Species in this category require somewhat larger openings than species in category 2 to achieve dominance, but, like those in category 2, they can respond to the whole range of larger openings.
4. *Upgrowth by seedlings and sprouts of shade-intolerant species.* Individuals of this category are usually not found under the closed canopy of the northern hardwood forest but in well-lighted openings. Tree species in this category require larger openings than those of category 3. Generally, openings $>400\text{-m}^2$ are required before these species contribute to the subsequent dominance of the opening. However, raspberry, a thoroughly shade-intolerant shrub, is often found in relatively smaller openings. Species in category 4 respond with increasing vigor as the size of the opening increases. Large openings often involve considerable soil disturbance, which contributes to the establishment of shade-intolerant species of the fugitive type (Table 4-1).

## COMPOSITION OF THE DOMINANT LAYER DURING ECOSYSTEM DEVELOPMENT AFTER CLEAR-CUTTING

During the first several decades after cutting, as the canopy closes, canopy differentiation resulting from competition (Figure 4-10) among established even-aged trees is the most important mechanism determining the structure of the dominant layer of the ecosystem. Endogenous disturbances are relatively unimportant, since dying trees are only of modest size, and vacated space is filled by individuals whose origin dates to the time of cutting. This natural thinning process results in increasing dominance of the overstory and understory by even-aged shade-tolerant and intermediate species which have the potential to remain alive for a long period of time. Shorter-lived intolerant and intermediate species are eliminated or greatly reduced in importance in the canopy. Because of the continuous canopy and modest endogenous disturbances, intolerant and intermediate herbaceous and shrub species are eliminated or markedly reduced as components of the active living biomass. They may persist, however, as dormant living biomass, e.g., dormant seeds or vegetative structures.

For the first century, the general trend of differentiation, acting on the pool of individuals established immediately after clear-cutting, is as follows:

1. development of a largely even-aged overstory increasingly dominated by long-lived shade-tolerant and intermediate species;
2. decrease in diversity of higher plants, with all layers of the ecosystem increasingly dominated by shade-tolerant species;
3. decrease in the density of dominants, coupled with an increase in average biomass;
4. continued biomass accumulation by the ecosystem (Figure 2-1).

However, as tree size increases larger endogenous disturbances result. These holes in the canopy are at first modest in size, perhaps 50 to 200 $m^2$, as trees of intermediate size topple, and tend to be filled principally by upgrowth of shade-tolerant species (category 2; Figure 4-10) that had become established on the forest floor during the initial period of canopy differentiation. This process, discussed by Forcier (1973), contributes to a further increase in the importance of tolerant species in the mid-aggrading ecosystem.

With time, some of the dominant trees begin to achieve really massive sizes, up to $>8$ t aboveground fresh weight (Figure 4-9). The subsequent fall of these trees creates larger openings, in the range of 300 to 500 $m^2$. These openings favor not only subsequent dominance by advance-regeneration of shade-tolerant species but also upgrowth by intermediate species and, occasionally, upgrowth of shade-intolerant species (Figure 4-10). Local environmental enrichment is greater, and consequently selection is somewhat more favorable to species employing the exploitive strategy.

Data on stand development (see page 115) suggest that, given full stocking by a diversity of species after clear-cutting and no exogenous disturbance, subsequent differentiation of the vegetation will first occur by natural thinning, with development of an even-aged overstory increasingly composed of long-lived shade-tolerant and intermediate species as thinning proceeds. With time, however, the magnitude of endogenous disturbance increases with increasing tree size, and this promotes the gradual development of an all-aged overstory as younger individuals from the forest floor grow into the canopy. At first, upgrowth of new stems is largely by shade-tolerant species and contributes to the increasing importance of these species in the ecosystem. Eventually, as trees age and grow larger, the potential for endogenous disturbance grows, to the point at which upgrowth by intermediate species is not uncommon, and even shade-intolerant trees have the opportunity of growing into the canopy from the forest floor.

Shade-intolerant species are usually considered early-successional species. However, our data suggest that even without exogenous disturbance (e.g., cutting, fire, and windthrow) the ecosystem contains a potential for endogenous disturbance sufficient to maintain intermediate species as well as intolerant species as part of the old, uneven-aged forest. Given the weight and disturbance potential of large trees, there seems to be no way to exclude intolerant species from the old-aged northern hardwood ecosystem undisturbed by exogenous forces. Situations where intolerant species are wholly absent are probably middle-aged developmental sequences with a temporary absence of intolerant species rather than so-called "climax" stands.

## SPECIES RICHNESS

An interesting trend in species richness (the number of species per square meter of land surface) seems evident during our hypothetical successional sequence. Species richness apparently would be at a maximum 2 or 3 years after clear-cutting (Figure 4-3), when the vegetation forms a relatively undifferentiated layer and contains a variety of life-forms and tolerance classes (Table 4-5). If we considered species richness by separate layers, (i.e., overstory, understory, and shrub and herb) with respect to time, we would expect species richness of upper layers to decrease with time because, with the development of stratification, various life-forms would be restricted to particular layers. However, once the upper canopy closes and the process of differentiation begins (see page 115), there would be a trend toward the elimination of intolerant and some intermediate species in all layers. It seems likely that species richness would reach a low point somewhere in the mid-Aggradation Phase, when the canopy is fairly continuous and dominated by tolerant and intermediate species producing heavy shade. Lower layers might be augmented by a few tolerant species, but, in general, herb cover would be modest and dominated by relatively few species (Siccama et al., 1970). Beyond that point, we would expect species richness to increase as the endogenous disturbance created by the fall of large trees becomes a progressively more important factor. We would not expect species richness in the older aggrading forest to equal that found shortly after a clear-cut, because richness is at a maximum in the most disturbed areas, and these constitute only a relatively limited part of the old aggrading forest, while they make up the entire area of a recent clear-cut.

Sprugel (1976), working with high-elevation fir forests of the northeastern United States has reported approximately the same pattern. Species richness and equitability (the Shannon–Wiener $H'$) reach a high peak immediately after disturbance, decline steeply to a low point as vigorous young forest develops, and once again increase to a middle value as the forest matures and opens.

It is interesting to note that in both of these examples maximum diversity is associated with maximum disturbance, increased resource availability, and minimum ecosystem stability (judged by ecosystem control over exports of water and nutrients). Conversely, minimum diversity is associated with those periods of time in which ecosystem control over exports is maximal, i.e., the mid-Aggradation Phase.

## Summary

1. The northern hardwood ecosystem contains species with a wide array of growth and reproductive strategies.
2. Clear-cutting activates reproductive strategies. The vegetation immediately after cutting is often a mixture of advance regeneration and new individuals that have arisen by sprouting or germination of seeds already present plus those transported to the site.
   a. The buried-seed strategy is particularly significant in revegetation in the Hubbard Brook region. This strategy involves:
      i. large quantities of dormant seed in the soil;
      ii. germination triggered by environmental signals associated with disturbance;
      iii. rapid height growth;
      iv. relatively rapid sexual reproduction with reestablishment of the dormant seed pool.
3. Clear-cutting results in populations of trees of different species but of about the same age. For a fairly long period, the forest is dominated by relays of even-aged trees as the ecosystem develops.
   a. Dominance of trees in the overstory changes with time as a result of different growth strategies of the tree species established immediately after clear-cutting.
   b. Dominance shifts from shorter-lived, faster-growing, less-tolerant species, which utilize an exploitive growth strategy, to longer-lived, slower-growing, tolerant species, which utilize a conservative but more-competitive growth strategy.
   c. Components of exploitive and conservative strategies are discussed.
4. Individuals starting growth at the time of clear-cutting have the potential of dominating the site for more than a century, but as the Aggradation Phase proceeds endogenous disturbance caused by treefall becomes an important factor in determining dominance in the overstory.
   a. At first, endogenous disturbance favors reinforcement of shade-tolerant species in the overstory. However, as tree size increases and local disturbance resulting from treefall becomes larger, intermediate and even some intolerant species become more abundant in the overstory.

   b. Ecosystem development without exogenous disturbance leads to a forest composition that contains the whole range of northern hardwood species, including the so-called early-successional species.
5. Shortly after clear-cutting, when the ecosystem is least stable in terms of its control of hydrologic, biogeochemical, and radiant energy flux, species richness is apparently at a maximum. In contrast, during the mid-Aggradation Phase (about 80 years after cutting), when the ecosystem has most control over flux rates, species richness is comparatively low.

CHAPTER 5

# Reorganization: Recovery of Biotic Regulation

The aggrading northern hardwood ecosystem has a considerable capacity to exercise control over hydrology and biogeochemistry and to regulate the flow and use of solar energy (Chapter 2). When this control is at its maximum, the ecosystem is most stable, with highly predictable and low net losses of nutrients and a fairly constant annual evapotranspiration rate. Regulation is rooted in internal ecosystem processes such as transpiration, nutrient uptake, decomposition, mineralization, and nitrification. Clear-cutting results in marked changes in internal ecosystem processes and a distinct loss of biotic regulation over energy flow, biogeochemistry, and hydrology (Chapter 3).

We now examine the recovery of biotic regulation as the various species respond, through growth and reproductive strategies, to the new conditions of resource availability created by clear-cutting.

Loss of regulation is reflected in markedly increased exports of water and nutrients from the ecosystem and some increase in soil erosion, depending on the degree of disturbance. On the other hand, recovery of the ecosystem may be stimulated by temporary enrichment of the site with available soil moisture during the summer, increased radiant energy at the ground level, and a marked increase in the concentrations of dissolved nutrients in the soil solution (Figure 5-1). In a sense, the ecosystem has an abundance of solar energy, is fertilized with nutrients drawn out of the nutrient capital of the system and is irrigated.

In the remainder of this chapter, we shall consider how the autotrophic biota respond to these conditions of site enrichment and how the ecosystem recovers biotic regulation over export patterns. In the course

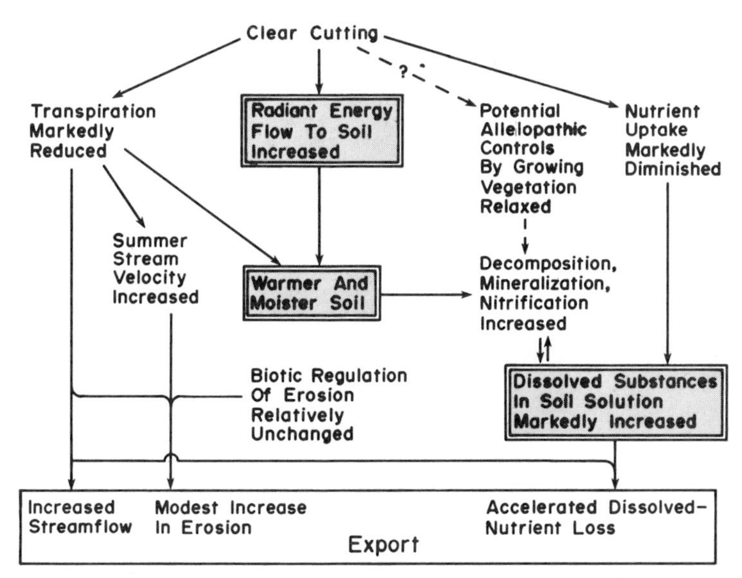


**Figure 5-1.** Hydrologic and biogeochemical responses of the northern hardwood ecosystem during the first two seasons after cutting. Erosion response assumes little damage to the soil during the harvesting process. Items in double-lined boxes represent marked increases in resource availability within the ecosystem.

of this discussion, we shall attempt to illustrate the array of mechanisms by which the short- and long-term stability of the ecosystem is maintained, the cost of such maintenance, and to present a generalized view of ecosystem homeostasis.

## PRIMARY PRODUCTIVITY

### Initial Response

Conditions created by clear-cutting activate the entire array of reproductive and growth strategies discussed in Chapter 4, and the clear-cut site is soon covered with vegetation. However, these general observations leave several basic questions unanswered. How rapidly does primary production recover? How does production compare to that prior to disturbance? How closely is production coupled to the heterotrophic processes discussed in Chapter 3? How quickly does the ecosystem reestablish maximum biotic regulation over energy flow, biogeochemistry, and hydrology?

Studies in the vicinity of Hubbard Brook (Figure 5-2) indicate that, after a lag of about 1 year, recovery of net primary productivity (NPP) on clear-cut sites is rapid. Within 2 years after cutting, NPP averages about 38% of a 55-yr-old forest. Leaf area index (LAI) is about 3.0 for 2-yr-old sites dominated by raspberry (Whitney, 1978), as compared to a value of about 6.0 for the older forest (Whittaker et al., 1974). At 4 years, NPP and LAI are about equal to values for a 55-yr-old forest (Figure 5-2; Marks, 1974). Between Years 4 and 6, NPP may rise substantially above that of the intact forest, but whether it does so appears to be influenced by the relative importance of exploitive species in the revegetation process. All of the points above the regression line in Figure 5-2 represent data from plots almost wholly dominated by exploitive species (raspberry or pin cherry).

The rapid average rate of NPP increase, 232 g/m$^2$-yr (Figure 5-2),

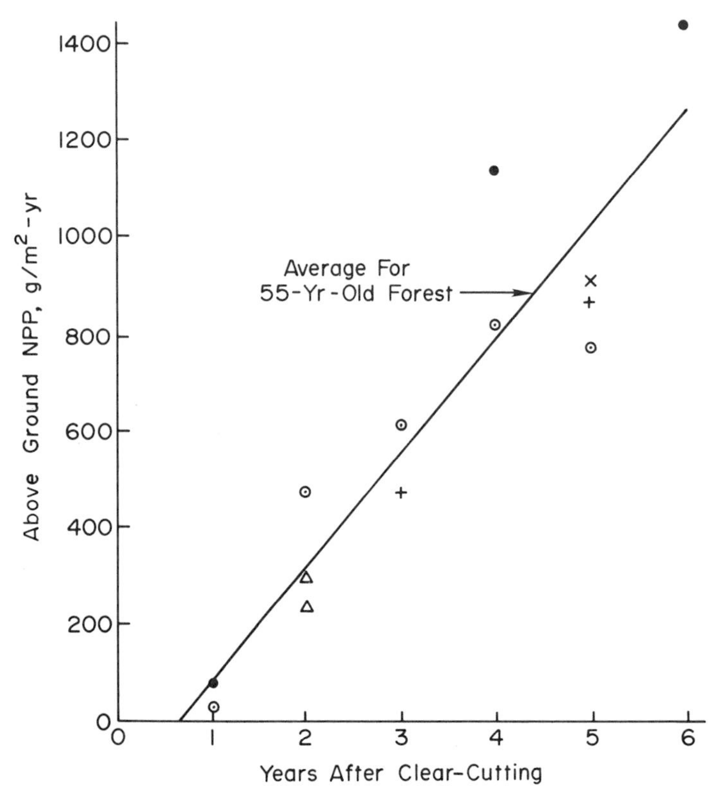


**Figure 5-2.** The relationship of aboveground net primary productivity to time after clear-cutting. Data are from the following sources: (●) Marks (1974); (⊙) Whitney (1978); (+) S. Bicknell, F. H. Bormann, and P. Marks (unpublished data); (△) P. Murphy (unpublished); (×) D. Ryan (unpublished). Murphy's data are for aboveground current growth and hence are a low estimate of aboveground NPP. All data are from plots where the forest floor had not been massively disturbed in the harvesting process. Safford and Filip (1974) suggest much lower rates of NPP on 4-yr-old plots that were scarified after cutting with a crawler tractor and a rock rake. It seems likely on clear-cut sites with much surface disturbance that average rates of NPP recovery would be lower than those presented here.

during the first 6 years indicates a strong autotrophic growth response to the increase in resource availability resulting from the clear-cut. For the ecosystem as a whole, during the course of development, it seems probable that there are few sustained rates of NPP increase that remotely approach that of the first few years after clear-cutting. The rapid rate of NPP increase and the relatively high absolute rates of primary production within 3 or 4 years after cutting suggest a fairly rapid response of autotrophs to increased resources resulting from heterotrophic processes (Chapter 3).

## Longer-Term Response

We have relatively few data on the productive behavior of the northern hardwood ecosystem after the initial burst of growth that follows clear-cutting. Obviously, NPP cannot increase indefinitely, nor does a single asymptote stretching over 170 years seem likely. According to our best understanding, the ecosystem undergoes a variety of biotic and abiotic changes during this long period, and it seems likely that these changes are reflected in primary production. There are some data to support this contention. Marks (1974) reports aboveground NPP of 1076 $g/m^2$-yr for a 14-yr-old pin cherry stand, while values of 957 and 792 $g/m^2$-yr were measured for a 40–45- and 45–50-yr-old stand dominated by sugar maple, beech, and yellow birch (Table 1-1). These data suggest that net primary productivity during the Reorganization Phase peaks and then declines.

Studies by Covington and Aber (1979) suggest a pattern for primary productivity during the Reorganization and Aggradation Phases of northern hardwoods. They studied leaf production in stands of different ages that arose naturally after clear-cutting and which exist within the limits set forth in Chapter 1. To these data they added appropriate leaf-production values from other studies. Leaf production is not a good index of changes in NPP during succession because of the changing proportions of NPP invested in photosynthetic and respiratory tissues as development proceeds (Whittaker and Woodwell, 1967). Leaf production, however, may be considered as a rough index of gross primary production (GPP). The curve generated by Covington and Aber (Figure 5-3) thus provides an index of changes in GPP through time. Although the fit of the curve is largely speculative, there is an underlying logic to it.

The curve suggests a rapid rise in GPP during the earliest years after clear-cutting. This conforms with the NPP measurements discussed above and can be accounted for by the high levels of resource availability resulting from the cutting.

There is some question as to how to draw the curve beyond Year 6. Should it gradually rise to a maximum at about Year 60, or dip and then rise, as Aber and Covington have drawn it? Use of data from different stands could be misleading; one stand may be simply more productive or

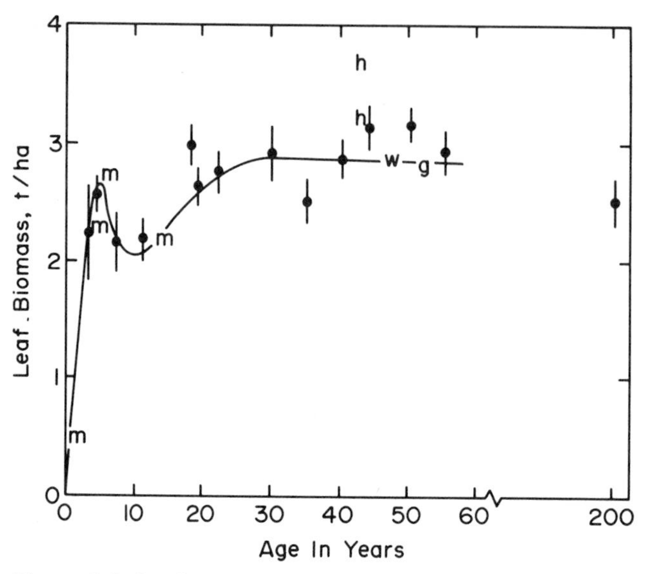


**Figure 5-3.** Leaf biomass (●) for a secondary developmental sequence in northern hardwoods. Age is years after clear-cutting. Data are means and 95% confidence intervals. Additional data are from (h) Hart et al., 1962; (g) Gosz et al., 1972; (m) Marks, 1974; and (w) Whittaker et al., 1974 (Covington and Aber, 1978).

variance may be such that a dip appears because data points are lacking. However, several lines of reasoning suggest that the dip and subsequent rise to Year 60 may be justified.

The proposed dip in production could result from the interaction of many factors. First, reduction in the amount of easily decomposable organic substrate in the forest floor could diminish nutrient availability (Chapter 3). The decline in depth of the forest floor means less mineralization and also less storage space for water and available nutrients. Second, the rapid development of high rates of GPP shortly after cutting implies increasing competition for water and nutrients, which might differentially affect productivity of exploitive species since they are probably less able to compete for water and nutrients at low ranges of availability. Third, Covington and Aber point out that the transfer of productivity from one species to another may not be smooth, and a temporary drop in production may result. For example, toward the end of the Reorganization Phase, as pin cherry begins to die out, the uptake of available resources may be diminished, and for a time the other species may not take up the slack.

Explanation of the relatively rapid rise after the decrease in production is even more problematical. Yet the proposed rise in GPP to about Year 50 (Figure 5-3) appears reasonable, since other measurements indicate that the ecosystem is simultaneously undergoing a rapid accumulation of net biomass in three of its four compartments: living organisms, forest floor, and dead wood (Figure 2-1). There are several factors that could contribute to a rising GPP. A shift in predominance from species utilizing the exploitive growth strategy to those utilizing a conservative growth strategy could lead to greater production through more efficient use of site resources. Secondly, initial growth seems to be primarily supported by nutrients drawn largely from the forest floor and the upper few centimeters of the mineral soil by shallow-rooted species. Gradual occupancy of the site by deeper-rooted species would allow utilization of nutrient and water resources deeper in the mineral soil. Although cation exchange per unit weight of soil in the B-horizons of White Mountain soils is low in comparison to that of the forest floor, the total amounts are substantial owing to greater thickness and bulk density. Similarly, more than 90% of the soil's water-holding capacity is found in the mineral soil (Hoyle, 1973).

The paucity of data on productivity beyond about Year 55 makes speculation difficult. However, in an old aggrading forest, dominated by large old trees, there would be a high ratio of respiratory to photosynthetic tissue (Whittaker and Woodwell, 1967), and this could lead to relatively low net primary productivity, since GPP minus green plant respiration equals NPP.

## RECOVERY OF BIOTIC REGULATION OVER ECOSYSTEM EXPORT

How quickly does the revegetating ecosystem establish biotic regulation over export patterns? We have sought answers to this question in several ways: by following the recovery of the deforested watershed (W2) at Hubbard Brook, by comparing stream-water chemistry from revegetating, commercially clear-cut forests and adjacent aggrading forests, and by drawing on information from the literature.

### Recovery After Enforced Devegetation

The sequence of revegetation of the deforested watershed (W2) at Hubbard Brook is shown in Figure 5-4. Once revegetation was allowed to occur, the rate of recovery of net primary productivity was initially much lower (Reiners et al., 1979) than in areas where natural revegetation immediately followed clear-cutting (Figure 5-5). On W2 there was a 2-yr period of very low growth, as compared to a single year of low growth on the commercially cut watersheds. NPP increased more or less linearly after the second year of recovery on W2, but the rate of increase from Year 2 to Year 5 was about 27% lower than that of immediate-growth systems. By Year 5, average aboveground NPP on immediate-growth systems had risen to over 1000 $g/m^2$, while on W2 it was about 45% lower.

(A)

(B)

**Figure 5-4.** The revegetation of Watershed 2. (A) June 1968, the beginning of the last of three growing seasons when the watershed was kept bare by herbicide treatment. (B) June 1970, the beginning of the second growing season of revegetation.

(C)

(D)

**Figure 5-4.** (*cont.*) (C) September 1971, the end of the third growing season of revegetation. (D) Late May of 1972. This sequence of photographs illustrates the considerable capacity of the ecosystem to revegetate despite 3 consecutive years of enforced barrenness. Photos courtesy of the U.S. Forest Service.

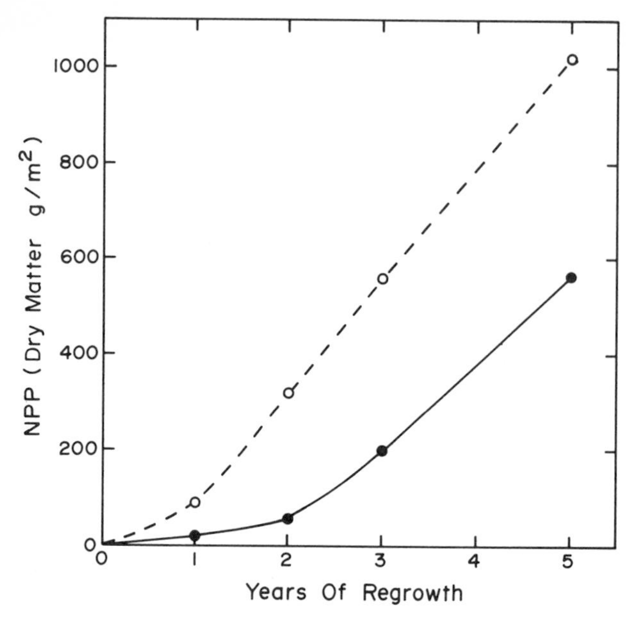


**Figure 5-5.** Aboveground regrowth (NPP) after (●—●) clear-cutting and enforced devegetation (W2, Reiners et al., 1979) and (○---○) clear-cutting followed immediately by natural revegetation (from Figure 5-2).

These fundamental differences in productive behavior plus 3 years of enforced devegetation suggest that biogeochemical data for the recovery period from W2 may not precisely portray the reestablishment of biotic regulation in commercially clear-cut systems that are allowed to revegetate immediately. Nevertheless, hydrologic and biogeochemical data from W2 are the most complete in existence, and they do reveal insights into basic patterns of ecosystem recovery.

Annual streamflow on W2 increased between 24 and 35-cm during the devegetation period. These values represent 26 to 41% greater streamflow than would have occurred if the forest had been left uncut (Chapter 3; Figure 5-6). Variation in streamflow during this period was due to differences in amount and timing of precipitation (Hornbeck, 1975). Almost all of these increases occurred during the growing season and were coincident with the elimination of transpiration by cutting. Streamflow decreased markedly during the first 4 years of revegetation (Figure 5-6). Although much of that decrease was probably due to increases in transpiration related to revegetation, some proportion was apparently related to chance variations in amount of precipitation, since streamflow rose again during the last 3 water-years reported here, 1974–75 to 1976–77. Thus after 7 years, despite moderately high levels of primary productivity, the export of liquid water remained significantly higher than that expected from a 60-yr-old aggrading forest (J. Hornbeck, personal communication).

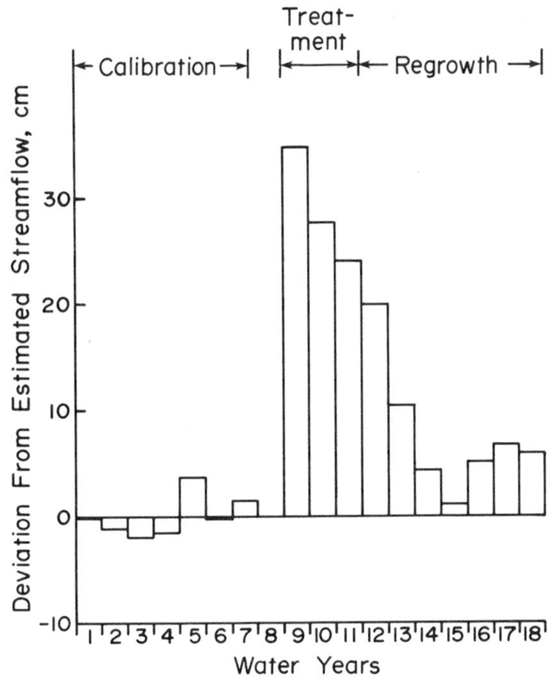


**Figure 5-6.** Deviations from regression of annual streamflow for Watershed 2 on annual streamflow for adjacent uncut reference Watershed 3. Period A represents precutting calibration period; Period B is the enforced devegetation period, and Period C is the regrowth period. Water-years are from 1 June to 31 May and begin 1 June 1958 (from Hornbeck, 1975, and J. Hornbeck, personal communication).

Concentrations of dissolved substances (Table 5-1) and their export from W2 (Figure 5-7) decreased markedly during the first 3 years of the recovery period. The explanation for this decrease is not entirely clear, but nutrient uptake by vegetation was not solely responsible. The decline in annual concentrations of nitrate, potassium, and calcium in stream water began during the third year of devegetation in the absence of any vegetation. Earlier (Chapter 3) we suggested this might be due to the exhaustion of easily decomposable substrates. A modest decline in nutrient concentrations in stream water was recorded during the first year of revegetation, followed by a sharp decline in the second and third year. Primary production during the second year was relatively low, yet the decreases in stream-water concentrations were among the sharpest recorded (Table 5-1). Several interpretations are possible, for example: (1) exhaustion of decomposable substrate may have continued to be a major factor; (2) relatively small amounts of vegetation growing along drainage channels may have had a major effect in determining stream-water chemistry; (3) nitrification may have been inhibited by living plants. By the third year of the regrowth period (Table 5-1), nitrate concen-

**Table 5-1.** Annual Weighed Concentrations of Nitrate, Calcium, and Potassium Ions in Stream Water from Reference Watershed 6 (W6) and the Experimentally Devegetated Watershed 2 (W2)[a]

| | Nitrate | | Calcium | | Potassium | |
|---|---|---|---|---|---|---|
| Year | W6 | W2 | W6 | W2 | W6 | W2 |
| *Precutting period* | | | | | | |
| 1963–1964 | —[b] | —[b] | 1.5 | 2.1 | 0.3 | 0.3 |
| 1964–1965 | 1.0 | 1.3 | 1.0 | 1.4 | 0.2 | 0.2 |
| 1965–1966 | 0.9 | 0.9 | 1.4 | 1.8 | 0.2 | 0.2 |
| *Devegetated period* | | | | | | |
| 1966–1967 | 0.7 | 38.4 | 1.3 | 6.5 | 0.2 | 1.9 |
| 1967–1968 | 1.3 | 52.9 | 1.3 | 7.6 | 0.3 | 3.0 |
| 1968–1969 | 1.3 | 40.4 | 1.3 | 6.0 | 0.3 | 2.9 |
| *Regrowth period* | | | | | | |
| 1969–1970 | 3.2 | 37.6 | 1.6 | 6.2 | 0.3 | 2.8 |
| 1970–1971 | 2.4 | 18.8 | 1.5 | 4.0 | 0.2 | 1.6 |
| 1971–1972 | 2.3 | 2.7 | 1.4 | 2.2 | 0.2 | 0.7 |
| 1972–1973 | 1.8 | 0.3 | 1.4 | 1.8 | 0.2 | 0.5 |
| 1973–1974 | 2.3 | 0.1 | 1.4 | 1.7 | 0.2 | 0.5 |
| 1974–1975 | 2.4 | 0.1 | 1.4 | 1.7 | 0.2 | 0.4 |
| 1975–1976 | 2.2 | 0.1 | 1.3 | 1.6 | 0.2 | 0.4 |

[a] *Values are in milligrams per liter.*
[b] *No data.*

trations were indistinguishable from those of an approximately 60-yr-old aggrading system, and beyond that time they were lower. Calcium concentrations after the third year of regrowth were about the same as those found in the 60-yr-old forest, but potassium concentrations remained relatively high.

Erosion and transportation of particulate matter were very responsive to revegetation (Figure 5-7). Toward the end of the devegetation period and during the first year of vegetation regrowth, the export of particulate matter was increasing exponentially. This trend was reversed during the second and third year of recovery; erosion decreased sharply, coincident with a relatively small amount of vegetation growth. With such small amounts of primary production (Figure 5-4), regulation of particulate export obviously did not come from reestablishment of the forest floor through litterfall. Observations suggest that a concentration of plant growth in and around stream banks might have been responsible for the stabilization of stream banks and reestablishment of organic-debris dams. Stream banks, by virtue of direct exposure to streamflow, are more subject to the erosive force of flowing water than any other single area of the ecosystem.

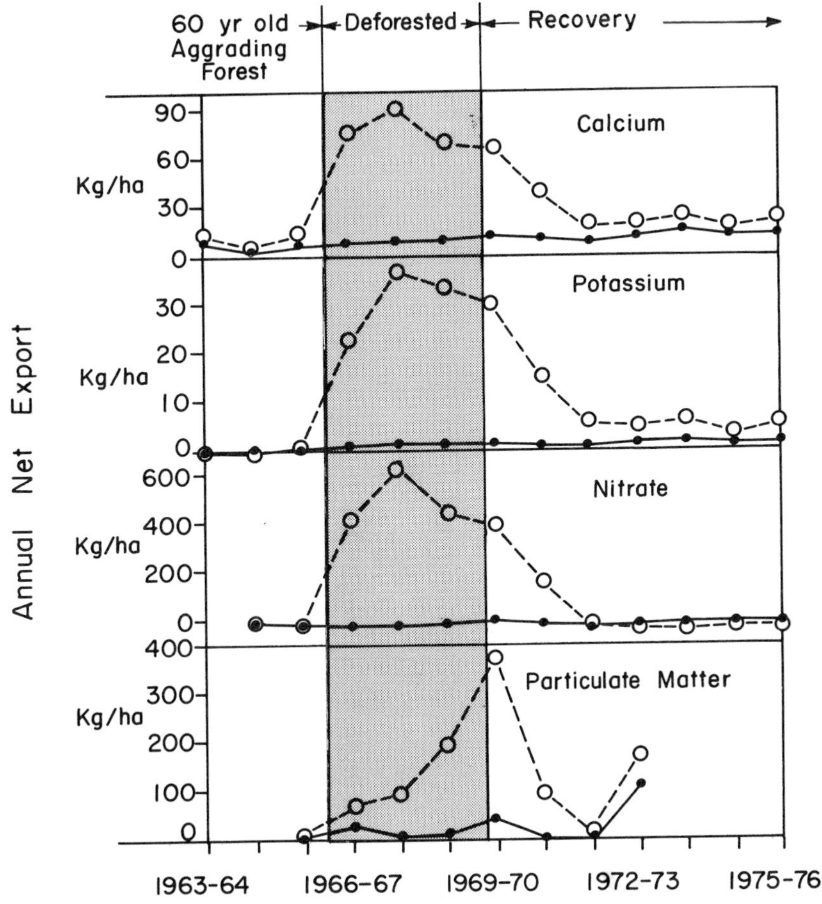


**Figure 5-7.** Export patterns of dissolved substances (calcium, potassium and nitrate) and particulate matter in stream water from (○--○) the experimentally clear-cut watershed (W2) and (●—●) the forested reference watershed (W6) (modified from Likens et al., 1978).

Information from W2 contributed in the following ways to our understanding of the recovery of biotic regulation of export patterns in clear-cut ecosystems:

1. The massive loss of nutrients from W2 resulted in part from the delay in coupling primary productivity and the hydrologic and nutrient enrichment that occurred immediately after disturbance. Not only did enforced devegetation prevent growth and reproductive strategies from taking effect immediately, but it reduced the effectiveness of these strategies once regrowth took place. The relatively sluggish growth once revegetation began was probably due, in part, to the loss of nutrients during the period of enforced devegetation.

These points emphasize the importance of effective and early coupling of forest floor decomposition and nutrient uptake by new vegetation. Any harvesting practice that tends to uncouple decomposition and nutrient uptake by disrupting the forest floor may contribute to unnecessary nutrient export and a reduction in the rate of recovery of primary productivity.

2. Despite 3 years of enforced devegetation and substantial loss of nutrients from the ecosystem, there were, coincident with revegetation, rapid initial declines in the concentrations of dissolved substances in stream water and the export of liquid water, dissolved substances, and particulate matter. Steep initial declines were associated with relatively small increases in vegetation growth, suggesting the possibility that living plants, particularly along drainage channels, have a great effect in establishing early regulation over export.
3. After 7 years of recovery, even though net primary productivity had risen to substantial levels (Figure 5-4), streamflow and the net export of calcium and potassium were still greater than precutting levels (Figures 5-6, 5-7). Nitrogen, on the other hand, after the fourth year of the regrowth period, showed a net gain (i.e., meteorologic input minus gross export gave a positive value), indicating that losses of dissolved nitrogen are rapidly brought under control by the revegetating system and that a net accumulation of nitrogen may have begun. Caution is advised here because we don't know the net balance for gaseous nitrogenous substances. These results indicate that each species of dissolved ion had a different recovery pattern (Likens et al., 1978).

## Recovery With Immediate Revegetation

Extensive data from throughout the eastern deciduous forest suggest that the export of particulate matter after clear-cutting is minimal if the site is rapidly revegetated and not excessively erosion-prone and if road construction and cutting are carried out in a careful manner (Patric, 1976b, 1977). Our findings for the northern hardwood forest are in accord with these data. In Chapter 2 we elaborated on a number of biotic mechanisms, relating to both living and dead biomass, that exercise regulation over the erodibility of the aggrading forest. Our experimental study of the devegetated ecosystem (W2) demonstrates that mechanisms controlling erodibility of the ecosystem can function at high levels for 2 years *in the absence of any green plant growth* as long as the forest floor is not excessively disturbed (Chapter 3). The recovery of the devegetated ecosystem indicates that relatively small amounts of vegetation regrowth can rapidly reduce export of particulate-matter. All of these factors taken together indicate that, with acceptable harvesting practices, biotic regulation of particulate-matter export after clear-cutting is only moderately diminished and recovers rapidly with reestablishment of vegetation.

Apparently partial biotic regulation of streamflow via transpiration begins rapidly with regrowth, but establishment of full regulation may take several decades. For example, the recovery of a clear-cut forest in North Carolina indicated a rapid initial decline in streamflow, but after 23 years streamflow was still singificantly greater than that expected if the watershed had not been cut (Figure 5-8). Similar data from West Virginia indicated that more than 10 years were required for streamflow to return to precutting levels (Lull and Reinhart, 1967). These data indicate that decline in streamflow is not a simple function of recovery of NPP. Other factors such as height and dispersion of the canopy (Douglass, 1967), percentage of canopy in evergreens (Swank and Douglass, 1974), percentage of soil occupied by roots (Lull and Reinhart, 1967), and changes in the water-holding capacity of the forest floor also may be determinants. In clear-cut northern hardwood ecosystems, it seems likely that recovery of full biotic regulation of streamflow takes 10 to 20 years, or to about the end of the proposed Reorganization Phase.

To determine the effect of immediate revegetation on stream-water chemistry, we made a comparative study of water from nine small commercially clear-cut watersheds and from nearby uncut stands (Pierce et al., 1972; three stands are not included here because one had an inadequate reference watershed, another was dominated by spruce and fir, and the third had a record of inadequate length). Stream-water

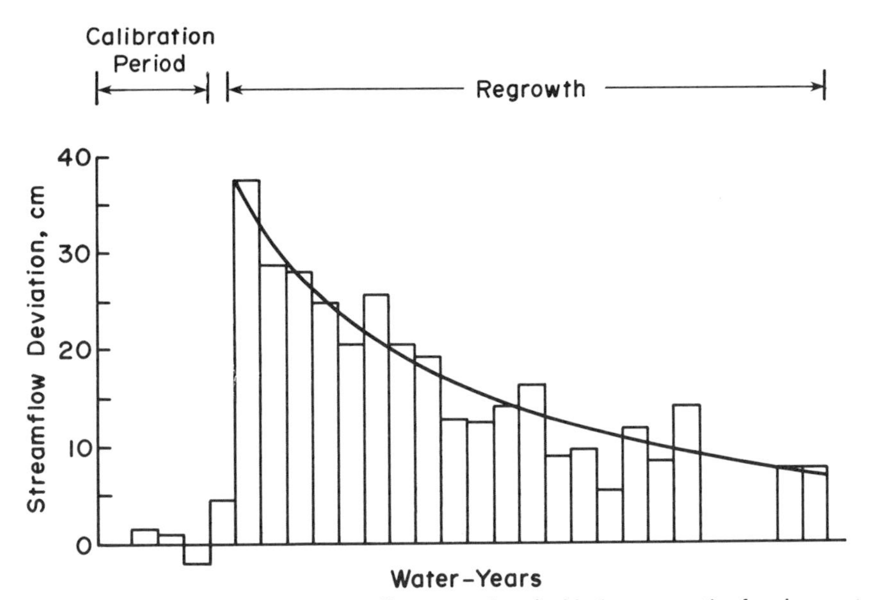


**Figure 5-8.** The decline in streamflow associated with the regrowth of a clear-cut forest at Coweeta, North Carolina. Decline in streamflow is calculated as deviation from expected streamflow. The zero line represents expected streamflow if the forest had not been cut. There is a 3-yr calibration period prior to cutting and a 23-yr period of regrowth (after Hibbert, 1967).

concentrations from the uncut stands were very similar to those reported for aggrading forests at Hubbard Brook (Chapter 2, Likens et al., 1977).

Although biotic regulation over stream-water chemistry was substantially weakened, it was not changed to the extent shown in W2 (the devegetated system, Table 5-2). For the six commercially clear-cut watersheds, annual concentrations (unweighted for volume) of nitrate-nitrogen, potassium, and calcium averaged for the first 2 years after cutting showed increases by factors of 12, 2, and 3, respectively (Table 5-2). This compares with increases by factors of 41, 11, and 5 for the devegetated watershed (W2) when compared to the average of all uncut watersheds at Hubbard Brook (Likens et al., 1970).

In contrast to the devegetated system at Hubbard Brook, only one commercially clear-cut watershed showed stream-water concentrations to be highest in the second year after cutting. Three commercially clear-cut watersheds had highest concentrations in the first year, and the pattern was not clear for the remaining two. In only two of the six watersheds did nitrate concentrations fall to precutting levels within 3 years after regrowth began, as happened on W2. Trends in the other four watersheds suggested that such a reduction would take 4 or 5 years after cutting. Calcium concentrations in two watersheds and potassium concentrations in five watersheds remained above reference levels throughout the measurement period of 4 years.

We draw two important conclusions from these results. Immediate revegetation reduces the magnitude of increases in concentrations of dissolved substances in drainage water, but it does not prevent significant increases from occurring. This indicates that coupling between mineralization processes (accelerated by clear-cutting) and nutrient uptake by regrowing vegetation is far from perfect and that other nutrient sinks within the ecosystem are incapable of retaining all nutrients not taken up by vegetation.

**Table 5-2.** Ratio of Stream-Water Concentrations of Commercially Clear-Cut Forests to that of Reference Streams Draining nearby Uncut Forests in the White Mountains of New Hampshire[a]

| Years After Cutting | No. of Sites Sampled | Clear-cut/Uncut Ratio | | |
|---|---|---|---|---|
| | | Nitrate | Calcium | Potassium |
| 1 | 4 | $14.0 \pm 5.0$ | $2.4 \pm 0.1$ | $3.0 \pm 0.7$ |
| 2 | 6 | $9.7 \pm 3.8$ | $2.1 \pm 0.4$ | $2.6 \pm 0.3$ |
| 3 | 6 | $3.8 \pm 1.6$ | $1.6 \pm 0.3$ | $2.1 \pm 0.3$ |
| 4 | 2 | $1.1 \pm 1.0$ | $1.4 \pm 0.4$ | $2.5 \pm 0.1$ |

[a] *Values expressed as means ± SEM. Ratios are based on unweighted averages (Pierce et al., 1972, and R. Pierce, W. Martin, and A. Federer, unpublished data).*

We conclude that reestablishment of biotic regulation over export of nutrients from commercially cut forests is similar to that shown for the experimentally devegetated system. Nutrient export in stream water is a function of concentration and amount of streamflow. In the northern hardwood system, regulation of stream-water concentrations is established fairly rapidly for nitrate (1 to 5 years) and more slowly for cations (5 to 10 years) and regulation of streamflow occurs still more slowly (10 to 20 years). Considering concentration and streamflow together, it is apparent that biotic regulation (over *large increases* in nutrient export) is achieved fairly rapidly, i.e., within 3 to 5 years after cutting, but that one or two decades is required before precutting levels of export are reestablished. Substantial quantities of nutrients are lost from the clear-cut ecosystem before full control is reestablished.

## COUPLING OF MINERALIZATION AND STORAGE PROCESSES

A major consideration in the analysis of ecosystem dynamics after disturbance is the efficiency with which the ecosystem is able to store nutrients. Storage may consist of the transfer of nutrients to new vegetation or to other sinks within the ecosystem. For example, following clear-cutting in a Douglas fir forest, a very different kind of ecosystem, little export of nutrients was reported, but major transfers from upper to lower horizons are thought to occur (Cole and Gessel, 1965; Gessel et al., 1973).

Covington (1976) has pointed out that during the Reorganization Phase more nitrogen is released from the forest floor than can be accounted for by a net accumulation in living biomass plus that lost as dissolved substances in stream water.

To gain a more detailed insight into mineralization–storage processes in northern hardwoods, we estimated nutrient losses from the forest floor, nutrient gains by regrowing vegetation, and net nutrient export of dissolved substances in stream water for an average stand over a 4-yr period after commercial clear-cutting (Figures 5-9, 5-10, and 5-11). These average data were obtained by compositing data collected in different stands at different times and thus should be considered as a first approximation. More precise data are needed, and a simultaneous study of all of these parameters should be conducted in a single stand over the same time period. Nevertheless, the present analysis suggests several interesting and to some degree startling conclusions.

1. In immediately revegetating clear-cut ecosystems, the growth response of new vegetation is not sufficient to prevent significant losses of dissolved substances. However, within a few years after cutting, annual increases of nutrients stored in biomass were approximately equal to or greater than annual *decreases* in stream-

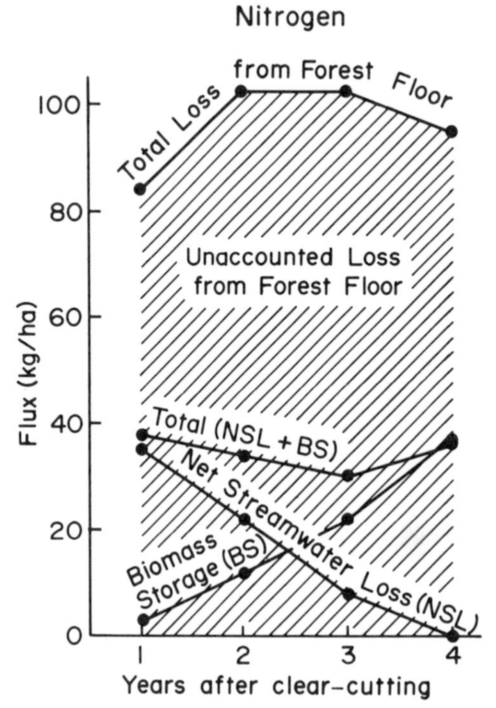


**Figure 5-9.** An estimated nitrogen budget for a commercially clear-cut forest. Net dissolved losses in stream water, storage in living biomass, and total losses from the forest floor are shown for 4 years after cutting. Forest floor data are for an average site and are estimated using nutrient concentrations and an equation for predicting changes in forest floor depth (Covington, 1976). Accumulation in living biomass is estimated from Marks (1974) and Likens and Bormann (1970), and the estimate is probably high. Stream-export data are calculated with the stream-water concentrations and estimated streamflow (Table 5-2) for six commercially clear-cut watersheds in the White Mountains. Unaccounted loss from the forest floor represents the difference between total loss (TL) from the forest floor and biomass storage (BS) plus net stream-water losses (NSL) i.e., [TL − (BS + NSL)].

water output of nutrients (Figures 5-9, 5-10, and 5-11). It would appear that with time after disturbance a direct relationship develops between nutrient storage in living biomass and stream-water chemistry. However, this relationship may not be simple or direct; for example, regrowing vegetation may take up $NH_4^+$ directly and hence limit the process of nitrification, which in turn affects cation exchange, and so forth (Chapter 3).

2. A large amount of nitrogen, presumably lost from the forest floor after clear-cutting, is not accounted for by stream-water loss or plant uptake and storage (Figure 5-9). If our estimate of the rate of nitrogen loss from the forest floor is correct, about 60% of the

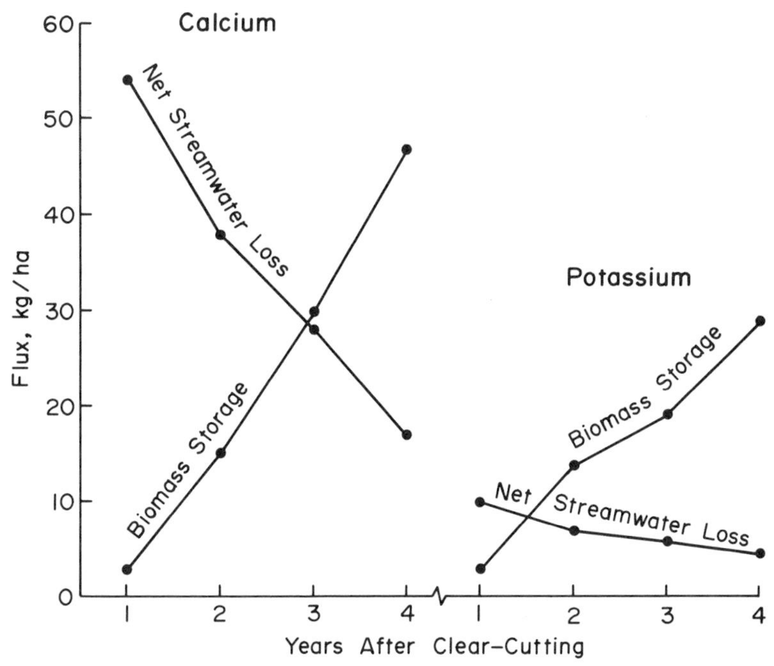


**Figure 5-10.** Estimated average biomass storage and net stream-water loss of potassium and calcium for a 4-yr period after commercial clear-cutting. See Figure 5-9 for methods of calculation.

predicted loss is unaccounted for by vegetation uptake and/or loss in drainage water. Covington (1976) suggests, for the Reorganization Phase as a whole, that some of the "missing" nitrogen may be tied up in decomposing logs left over from the clear-cutting. However, decomposing logs would be an unimportant "sink" during the first 4 years after cutting.

Nitrogen loss from the forest floor may be accounted for by translocation to deeper mineral horizons or volatilization by chemical or biological denitrification processes. The latter pathway would increase the already significant losses of nitrogen (i.e., dissolved nitrogen) from the ecosystem by a factor of five. It is interesting to note that Dominski (1971) also estimated nitrogen loss from the entire soil profile of W2 after 3 years of devegetation to be about 35% greater than stream-water losses. He suggested volatilization as an explanation, but his data were inconclusive.

These results indicate that there is much we do not know about the nitrogen cycle in forest ecosystems. The evaluation of denitrification under field conditions represents a major research challenge. It also should be borne in mind that a finding of significant denitrification would automatically entail still-greater amounts of

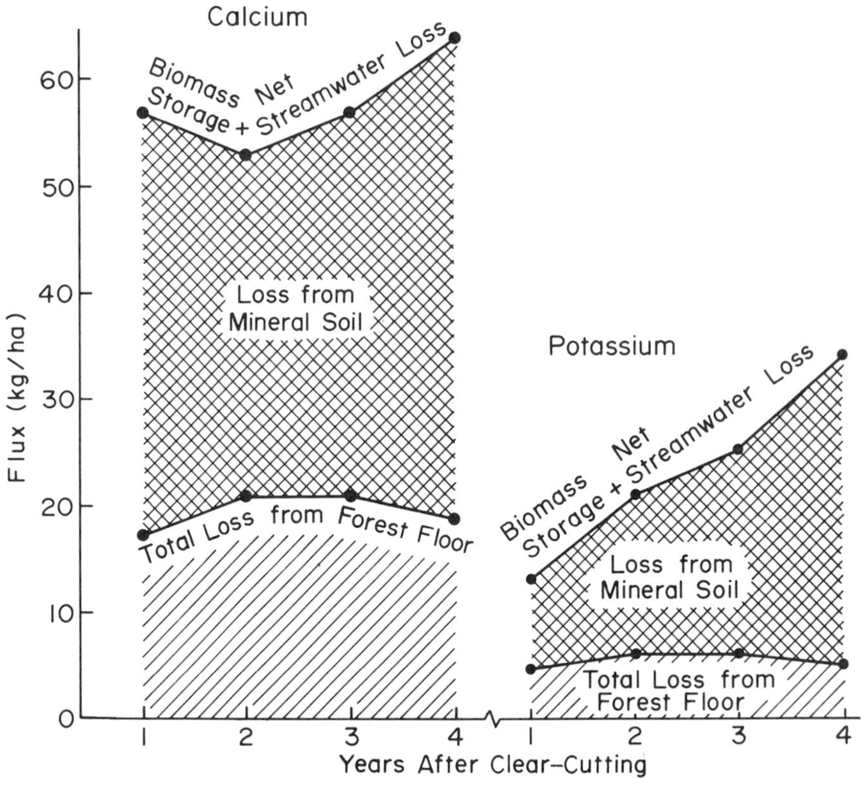


**Figure 5-11.** Flux of calcium and potassium in the ecosystem during the 4 years after commercial clear-cutting. Cations stored in biomass plus those flushed from the ecosystem cannot be wholly accounted for by losses from the forest floor. It is assumed the difference comes from the mineral soil. See Figure 5-9 for methods of calculation.

biological nitrogen fixation during the Aggradation Phase, to make up the additional losses.

3. Estimated losses of calcium and potassium from the forest floor cannot account for that taken up and stored in living biomass plus that lost in stream water (Figures 5-10 and 5-11). This suggests that cations are being withdrawn from the mineral soil as well as from the forest floor. Most of the calcium and potassium lost from the forest floor may be taken up by the shallow-rooted vegetation that typifies early revegetation (Table 4-7), while most of the dissolved ions exported from the ecosystem may be drawn from the deeper mineral soil. This is not an unreasonable assumption, since water filtering through the lower horizons is probably enriched in hydrogen ions resulting from the accelerated nitrification that follows clear-cutting, while lower horizons contain adequate exchangeable cations (Hoyle, 1973).

This discussion on nitrogen and cations allows us to comment on the behavior of the mineral-soil compartment during ecosystem development. In Chapter 1, we assumed that the organic content of the mineral soil remained unchanged during the development of the ecosystem after clear-cutting. We made that assumption because of our inability to measure changes accurately. The small amount of direct data we have (Dominski, 1971) suggests that the mineral soil loses organic matter and nitrogen immediately after clear-cutting, while the large amount of unaccounted-for nitrogen loss (Figure 5-9) suggests the possibility that the mineral soil may accumulate some nitrogen from the forest floor. These conflicting data reinforce our original decision to consider organic matter in the mineral soil as relatively unchanging. New research must be done to bridge this important gap in our knowledge.

The calcium and potassium data (Figure 5-11), on the other hand, seem more definitive; these data suggest it is unlikely that there was a net increase in exchangeable cations in the lower mineral soil after cutting of northern hardwoods such as that reported when Douglas fir was clear-cut (Cole and Gessel, 1965). Rather, the whole profile under northern hardwoods seems to lose exchangeable cations.

## REPLACEMENT OF LOST NUTRIENT CAPITAL

As discussed above, restoration of biotic regulation over hydrology and export of nutrients is not synchronous. Apparently, full biotic regulation over export is achieved 10 to 20 years after clear-cutting, roughly coincident with the end of the Reorganization Phase in our model of ecosystem development (Figure 1-7). Thereafter, the ecosystem acquires available nutrients from various meteorological inputs (including nitrogen fixation) and from weathering, as discussed in Chapter 2.

## ECOSYSTEM REGULATION

### The Aggradation Phase

*Biogeochemistry.* Our evidence indicates that various processes in the aggrading ecosystem are closely regulated. The most direct evidence comes from biogeochemical data. Of all the phases in the developmental sequence, the Aggradation Phase has the greatest degree of biogeochemical predictability and constancy. Despite highly variable precipitation chemistry (Chapter 2), concentrations of most dissolved substances in stream water have narrowly defined limits and are predictably related to flow rates or seasonal factors (Chapter 2). We find that the aggrading ecosystem is characterized by a relatively small range in amount of water vapor produced by evapotranspiration, despite a two-fold range in amount of precipitation (Figure 2-6), a small range in net export of dissolved substances, and highly controlled erosion with low export of eroded material (Chapter 2). We take constancy of export and biogeochemical predictability as evidence of a highly regulated system. The regulation of concentrations of dissolved substances is not determined simply by physical factors but results from the interactions between tightly coupled biotic and abiotic subsystems (Figure 3-5).

*Total Biomass Accumulation.* A striking feature of the Aggradation Phase of ecosystem development is the strong tendency to accumulate biomass; i.e., total biomass accumulation (living plus dead) occurs when gross primary productivity exceeds ecosystem respiration. The rate and amount of biomass accumulation is determined by the initial conditions of the site and the interactions between the biota and the external environment.

Regulation of the rate of biomass accumulation might be thought of as resulting from factors both external and internal to the ecosystem. External factors like solar energy, amount of precipitation, nutrient input, and temperature set theoretical limits on rate of biomass accumulation. If we consider external variables as constant, we might observe some maximum rate of biomass accumulation. In nature, however, internal factors such as changing species composition, competitive relationships, and insect defoliations cause the actual trend of biomass accumulation to depart significantly from any such theoretical maximum rate (Figure 5-12).

Changes in rate of biomass accumulation, whether it is due to a series

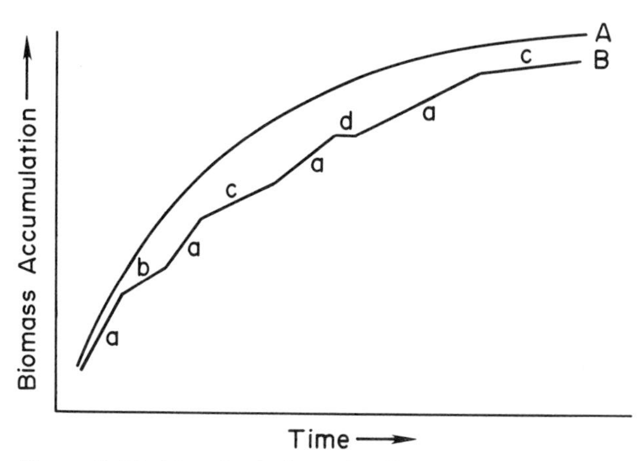


**Figure 5-12.** Hypothetical trends of biomass accumulation. (A) Trend of maximum potential biomass accumulation during the Aggradation Phase under constant and ideal growing conditions; the curve shows the decline in net biomass storage toward the end of the phase. (B) Trends in biomass accumulation as affected by the following departures from ideal conditions: (a) a series of good growing years, (b) drought periods, (c) a period of subnormal temperatures, and (d) an outbreak of defoliating insects.

of good or poor growing years, variable internal factors, or the inherent decrease with time in the storage of living and/or dead biomass within the ecosystem, must be reflected in the biogeochemistry of the ecosystem. Thus, even though our concept of ecosystem development suggests a highly predictable and fairly constant biogeochemistry during the Aggradation Phase, some variation is to be expected.

*The Relationship between Productivity and Resource Availability.* One of the factors that most affects the biogeochemical predictability and constancy of the Aggradation Phase is the generally close connection between primary productivity, resource availability, and nutrient regeneration. Thus, although major trends in GPP vary with time (Chapter 5) variations about the trend line itself tend to be relatively small. Aspects of this relationship are self-regulating; for example, competition among trees may contribute to the cybernetic regulation of primary productivity in a way analogous to that described for phytophagous insects by Mattson and Addy (1975). For example, strong competitors (intra- or interspecific) affect the physiological status of nearby but less competitive individuals, whose primary productivity is temporarily reduced. Demands of less competitive individuals on resource availability are relaxed; their resistance to predators and parasites is weakened and eventually they die. For a time, litterfall is increased, space in the canopy is surrendered, and resources previously used go unused. Canopy openings also may result in increased decomposition and nutrient regeneration. All of these responses contribute to increased availability of light, water, and nutrients to the surviving trees surrounding the opening. The stronger competitors respond with increased productivity, which is once again followed by increased competition, and the cycle is repeated.

If the sum of local increases in resource availability within the ecosystem was relatively great, while corresponding increases in productivity were relatively slow, one would expect increases in concentrations of nutrients in stream water, together with increased streamflow. That large oscillations in these biogeochemical parameters do not occur during the Aggradation Phase is probably related to the even-agedness of the aggrading forest. During the early part of the Aggradation Phase, competition between the largely even-aged trees results in death of many individuals, but gaps in the canopy are relatively small since trees are of small or modest size (Chapter 4). These small gaps probably have little affect on increasing decomposition rates and are rather rapidly closed by ingrowth of surrounding individuals. Thus, the feedback relationship (primary productivity $\rightleftarrows$ resource availability) does not swing through wide oscillations, and the more or less continuous canopy leads to close regulation of energy flow, hydrology, and biogeochemistry. Toward the end of the Aggradation Phase, competition and other factors that cause the death of trees result in larger gaps in the canopy, since the predominantly even-aged trees are now of larger size. These large gaps may foster locally increased decomposition and nitrification and marked local changes in productivity causing the equilibrium between productivity and resource availability for the ecosystem as a whole to exhibit somewhat wider oscillations. This in turn may result in slightly increased oscillations in the export of nutrients and water. The effect of declining biomass storage on biogeochemistry will be discussed in Chapter 6.

## The Reorganization Phase

The Aggradation Phase, with its fairly constant curve of biomass accumulation and its highly predictable biogeochemistry represents a system under strong regulation. In a sense it also represents a period of relative quiescence in our model of ecosystem development, a period with relatively little endogenous disturbance. In contrast, the Reorganization Phase begins with a major exogenous perturbation—clear-cutting. Clear-cutting sets in motion a series of unusually severe stresses whose initial effect is to reduce markedly the capacity of the ecosystem to regulate the flow of both radiant and mechanical energy, water, and nutrients in ways that diminish destructive effects. In a sense, the disturbance opens the system to those potentially degrading forces that are ever-present in its environment: rain, running water, wind, heat, and gravity. The longer these forces operate in an uncontrolled way, the more the ecosystem is degraded and the longer the time necessary for the ecosystem to recover to predisturbance conditions, if indeed recovery is possible.

For many humid terrestrial ecosystems, particularly those on slopes or with readily erodible substrates, erosion of particulate matter after disturbance presents great potential danger (Stone, 1973; Bormann et al., 1974). The physical removal of organic and inorganic materials not only results in loss of exchangeable cations attached to exchange surfaces but in the removal of the surfaces themselves. These losses in nutrient capital and exchange capacity in turn affect the productive capacity of the ecosystem and its ability to regain predisturbance levels of regulation over the flow of energy, water, and nutrients. In instances of severe erosion, an ecosystem might require a very long time to attain previous levels of production. Not only would productive capacity be sharply reduced from predisturbance levels, but redevelopment would involve the slow formation of new humic and clay surfaces and the sequestration of substantial amounts of nutrient capital. Unless checked by biological activities, accelerated erosion might continue until the ecosystem achieved a new relationship with the physical forces impinging on it; and this would occur at a much lower level of productivity and with far less ecosystem regulation of the physical environment (Figure 6-10).

*A Hypothesis of Homeostasis.* We propose that the severe stress initiated by clear-cutting not only accelerates the activity of some of the mechanisms responsible for biotic regulation during the Aggradation Phase but also calls into action another set of mechanisms largely quiescent during that phase. This idea is based, in part, on the set of relationships discussed in Chapter 3 and summarized in Figure 5-1.

Availability of many resources increases as a result of clear-cutting. There is a greater availability of soil moisture throughout the total profile caused by severely reduced transpiration. Removal of the canopy causes radiant energy reaching the forest floor to be increased by several orders of magnitude. Concentrations of dissolved substances in the soil solution are increased about an order of magnitude owing to accelerated decomposition and mineralization resulting from warmer and moister soil. There is also an increase in the activity of nitrifying organisms related to the removal of vegetation. Heterotrophic processes may for a time exhibit positive feedback (as shown in Figure 3-5) that raises levels of decomposition to a maximum the first year after cutting, while mineralization reaches its maximum during the second year.

The cutover ecosystem responds to these conditions of increased resource availability by a burst in primary production (Figure 5-2), not only by species that characterize the precutting forest but also by a group of species not part of the predisturbance forest (e.g., buried-seed species) that may have evolved specifically to fill a niche created by this type of disturbance (Chapter 4). Compared to the aggrading forest, primary productivity for the recently cutover ecosystem swings through at least one very wide oscillation. Cutting reduces productivity to near-zero, but productivity rapidly rises and for a few years may exceed that of the uncut aggrading forest. Thus, productivity is rapidly restored, and within a few years hydrologic and biogeochemical functions begin to approach the predisturbance levels discussed in Chapter 2.

In contrast to the rapid changes in available nutrients and water in the soil, increases in erodibility of the ecosystem are relatively slow to develop (Chapter 3). Processes related to dead biomass maintain a high level of regulation over erodibility for about 2 years. The clear-cutting and enforced-devegetation experiment showed that these mechanisms begin to weaken seriously only in the third growing season. In well-designed clear-cuts with immediate revegetation, accelerated productivity prevents serious erosion from ever occurring.

The coupling of heterotrophic processes to autotrophic processes is far from perfect. Considerable leakage of nutrients from the ecosystem occurs during the first few years, even in immediately revegetating clear-cut systems (Chapter 5). Such losses might be even higher if the release of gaseous nitrogen were included. This suggests that the rapid increase in productivity that follows disturbance is a relatively inefficient activity and is costly in terms of nutrient and biomass storage within the ecosystem. *However, the sacrifice of efficiency which results in accelerated production which may be considered an effective strategy of ecosystem stability since it forestalls a still-greater sacrifice in biotic regulation of erosion.*

*The Forest Floor as a Regulator of Ecosystem Activity.* The forest floor plays a major role in ecosystem homeostasis. Nutrient and energy costs during the critical period that immediately follows disturbance are paid by drawing on nutrients and energy stored in dead biomass, primarily in the forest floor. In a sense, the forest floor functions as a "regulator" for the ecosystem, with a net storage of energy and nutrients during the first third of the Aggradation Phase and a net liberation of energy and nutrients during the Reorganization Phase. This would be an important corollary to Margalef's dictum (1968) that "biomass is the keeper of organization," except that in this case dead biomass serves this function.

Not all northern hardwood ecosystems have a forest floor of the mor type common to northern New Hampshire. Some have a characteristic mull-type soil, where most leaf litter that falls on the mineral soil is rapidly decomposed, and the remainder is incorporated into the mineral soil (i.e., the $A_1$-horizon) by faunal activity (Lutz and Chandler, 1946; Bormann and Buell, 1964); mixing is due in large part to extraordinarily large populations of earthworms (Eaton and Chandler, 1942). Usually, mulls occur on richer sites and have a higher level of fertility than mor soils and are thought to sustain a high level of nitrification (Melillo, 1977). A basic question arises as to the behavior of organic matter in mull soils: Does it act as a regulator of ecosystem activity in the same way as organic matter in mor-type soils?

The large amount of dead wood on the forest floor immediately after a clear-cut may function as a nutrient reservoir in a manner somewhat similar to the forest floor. Dead wood is probably far less important during the critical years immediately after clear-cutting because of resistance to rapid decay. It seems likely that the principal importance of dead wood may be realized toward the middle and end of the Reorganization Phase, when the decomposition of slash would provide a modest source of nutrients for new growth. Perhaps of more importance, dead wood provides a substrate for the nitrogen-fixing organisms discussed in Chapter 2.

The role of dead wood in ecosystem dynamics needs careful study, particularly since new whole-tree harvesting techniques completely remove all aboveground wood from the clear-cut stand.

*Linkage between Primary Productivity and Heterotrophy.* The rapid response of the ecosystem to clear-cutting may be considered in terms of heterotrophic processes and primary productivity linked by a number of feedback channels. Perturbation favors an increase in heterotrophy, and, in fact, for a time various heterotrophic processes may be linked by positive feedback. Primary productivity is positively linked to heterotrophic end products as well as to soil moisture and energy conditions resulting from destruction of the original vegetation. As the rate of primary productivity increases, however, it exercises a negative-feedback control on heterotrophic processes. For example, we would expect available soil moisture to drop as primary productivity and transpiration, correspondingly, are increased and would expect nitrification to be limited as the vegetation reestablishes itself. Eventually, we would expect the rate of primary productivity to decline as the special growth conditions that immediately follow disturbance give way to conditions that characterize the intact aggrading system.

## SUMMARY

1. Primary productivity of northern hardwood ecosystems recovers rapidly after clear-cutting. This rapid increase in productivity is largely based on the growth response of exploitive species utilizing resources made available by destruction of the living vegetation and accelerated decomposition of the forest floor.
2. Biotic regulation of large losses of water and dissolved nutrients is established in 3 to 5 years after cutting, but losses may exceed those of uncut forests for 10 to 20 years, or to about the end of the Reorganization Phase.
3. Coupling between increased amounts of available nutrients after cutting and uptake of nutrients by regrowing vegetation is far from complete.
4. Comparison of nutrient uptake plus loss of nutrients in drainage water and estimated nutrient release from the forest floor suggests that a large amount of nitrogen is unaccounted for and that considerable cation leaching occurs from the mineral soil. Unaccounted for nitrogen loss from the forest floor may result from unmeasured denitrification losses or from transport to the mineral soil.
5. After a rapid early rise, primary productivity may decline at 6 or 7 years after cutting but recover and rise for several decades thereafter. The rise may be due to a shift from species utilizing inefficient exploitive growth strategies to those using more efficient conservative strategies, gradual occupancy and utilization of deeper layers of soil, and an increase in cation-exchange and water-holding capacity associated with the growth in mass of the forest floor.
6. The Aggradation Phase of ecosystem development is the most highly regulated of all phases in the developmental sequence. Regulation has a strong feedback component associated with the close relationship between primary productivity and resource availability resulting from the even-aged condition of the vegetation.
7. Clear-cutting initiates a strong homeostatic response based on feedback between factors controlling resource availability and those controlling primary productivity. This response brings exports of nutrients and water under control relatively quickly and prevents serious erosion from ever occurring.
8. Ecosystem recovery after disturbance is expensive and is powered primarily by energy and nutrients from the forest floor. The forest floor acts as a regulator of ecosystem activity, storing energy and nutrients in times of quiescence and releasing them in times of stress.

CHAPTER 6

# Ecosystem Development and the Steady State

The biomass accumulation model of ecosystem development which we propose (Chapter 1) has four phases after clear-cutting: Reorganization, Aggradation, Transition, and Steady State (Figure 1-2). In discussing this model, our strategy has been to move from the most verified to the least verified aspects. Many elements of the first two phases, Reorganization and Aggradation (Chapters 1–5), are based on observation and measurement of, and experiment with, actual stands; and the conclusions now can be tested and evaluated. We now consider the remaining phases of the model, Transition and Steady State.

The model illustrates one pathway by which an aggrading northern hardwood forest might develop into a steady-state ecosystem. *Our treatment is a theoretical one based on a logical assemblage of information on natural history and biogeochemistry, but it is not based on actual measurements of extant steady-state ecosystems.*

We propose an orderly pattern of autogenic development after clear-cutting, which, given sufficient time and a fairly constant macroenvironment, can lead to a steady-state condition. In nature, however, departures from this inherent tendency toward order are caused by fortuitous events (exogenous disturbances), which introduce an element of disorder into the pattern. These exogenous events complicate the interpretation of developmental relationships of the northern hardwood ecosystem. Thus, although it is easy to recognize extant ecosystems that correspond reasonably well to our proposed Reorganization and Aggradation Phases, it is difficult to find stands that correspond exactly to our proposed Transition and Steady-State Phases. Although there are a few candidates (e.g., Charcoal Hearth and Bowl Forests, New Hampshire), extant stands that correspond to the Transition and Steady-State Phases are difficult to identify with certainty. This results from several causes. Because of human activity, stands containing really old-aged trees are fairly rare, and their history of disturbance, or lack of it, is mostly unknown. Age of individuals in itself is insufficient to place a stand within a developmental sequence. Not only is age in northern hardwood species often difficult to determine because trees may have rotten centers, but old-aged trees can be found in the late Aggradation, Transition, or Steady-State Phases.

Despite the difficulty of verifying the proposed end point of ecosystem development, the model allows consideration of many aspects of development of northern hardwoods. Whether or not the Steady State is ever spatially significant is less important than understanding the mechanisms underlying the strong autogenic trend toward its achievement.

## EVIDENCE FOR A STEADY STATE

Quite apart from discussing the means by which a steady state is achieved, it is important to ask if the idea of Steady State itself is reasonable. Woodwell and Sparrow (1965) define the Steady State as a condition in which there is no net change in total biomass over time and annual ecosystem respiration approximately equals annual gross primary productivity. Changes in total biomass within the ecosystem are governed by relationships between the four major biomass compartments (green-plant living biomass, dead wood, forest floor, and organic matter in the mineral soil; Figure 1-5).

Both the forest floor (Covington, 1976) and dead wood (A. Brush, personal communication) compartments reach an equilibrium (i.e., input $\simeq$ output) about midway through the Aggradation Phase (Chapter 1). If we assume that organic matter in the mineral soil behaves in the same way (Chapters 1 and 5), a steady state according to the Woodwell–Sparrow definition would be realized if the green plant living biomass became stabilized. We think it likely that stabilization in the amount of living biomass might be achieved several centuries after clear-cutting.

The idea of the steady state is among the more controversial of ecological constructs. A steady state concept based on the premise that ecosystem respiration equals gross primary productivity should be considered at best approximate. In a strict sense, there can be no absolute steady state but only a system undergoing slow long-term change.

## STEADY-STATE MODELS

A widely accepted model of primary and secondary ecosystem development (Margalef, 1968; Odum, 1969; Whittaker and Woodwell, 1972; Whittaker et al., 1974; Whittaker, 1975) is characterized by an asymptotic curve of biomass accumulation culminating in a steady state or "climax" condition at maximum levels of biomass (Figure 6-1). Our reasoning, however, suggests that an asymptotic curve of biomass accumulation is insufficient to portray the developmental potential of the northern hardwood ecosystem following clear-cutting.

Our model of biomass accumulation (Figure 6-1) departs from the asymptotic model in several significant ways. The standing crop of total biomass after cutting declines (Reorganization Phase), steadily increases to a maximum (Aggradation Phase), erratically declines (Transition Phase), and finally stabilizes with somewhat irregular oscillations about a mean (Steady-State Phase). This developmental pattern reflects the initial even-aged condition of the cutover northern hardwood ecosystem and the gradual change to an all-aged condition, which we term the Shifting-Mosaic Steady State. The biomass accumulation pattern we propose and

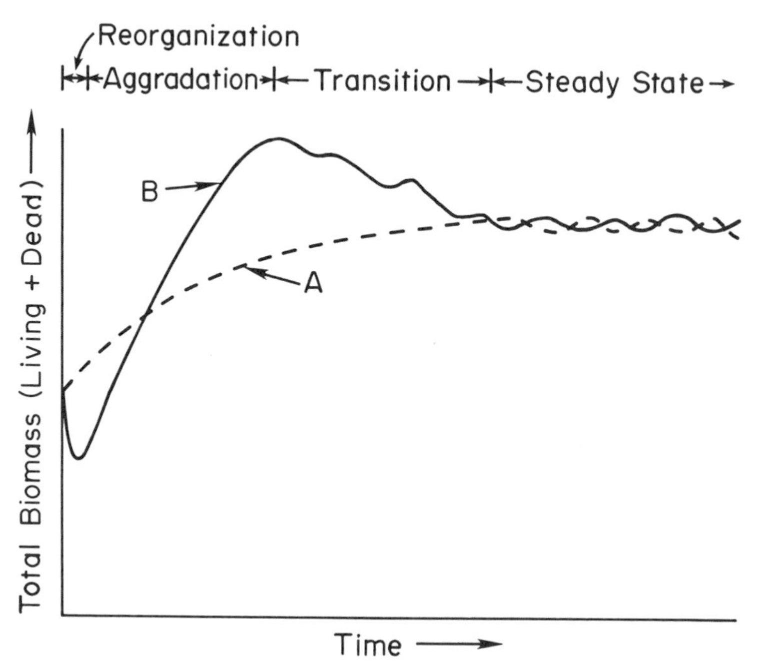


**Figure 6-1.** Two models of ecosystem development following clear-cutting (time zero) based on total (living plus dead) biomass accumulation. Line A is a simple asymptotic model that assumes a net biomass accumulation until the steady state is achieved and maximum biomass occurs during the Steady-State Phase; Line B is our model—which culminates in the Shifting-Mosaic Steady State. Developmental phases of our model are indicated.

the conditions that underlie its expression lead to conclusions about ecosystem energetics, structure, nutrient cycling, and stability quite different from those arising from the assumption of an asymptotic curve of biomass accumulation.

## LIVING BIOMASS ACCUMULATION

Our ideas about the behavior of living biomass accumulation through time were first derived by studying long-term simulations of forest growth generated with JABOWA, and, in the following paragraphs, we use JABOWA as a vehicle to elaborate our hypothesis on the development and nature of the steady state. We emphasize, however, that the logic underlying the hypothesis can stand on its own.

Long-term JABOWA simulations, beginning with a clear-cutting and calculated for constant environmental conditions, yield the same pattern of change in living biomass through time, i.e., rising to a peak, losing biomass, and then irregularly oscillating around some mean. Actual

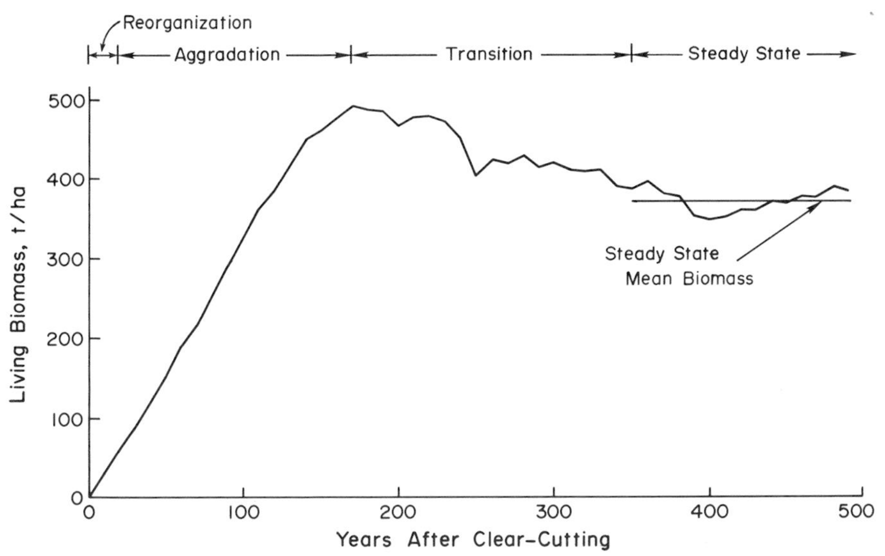


**Figure 6-2.** Changes in living biomass during ecosystem development after clear-cutting based on the average of two JABOWA simulations. Proposed phases of development (Reorganization, Aggradation, Transition, and Shifting-Mosaic Steady State) are shown. A curve of total ecosystem biomass (living plus dead) would tend to have the same pattern as for living biomass alone (Figure 6-1). This follows because (a) almost all input to the standing crop of dead biomass in both the forest floor and dead wood compartments is derived directly or indirectly from living plant biomass in the form of litter (Figure 1-5) and (b) standing crops of dead biomass in the forest floor and dead wood compartments tend to reach an equilibrium (Figures 1-8 and 1-9).

biomass values for the peak and the mean, which we define as steady state, vary with different runs as does the time necessary to descend from the peak biomass to steady-state biomass, but the same pattern is repeated every time. The curve of living biomass accumulation we show in Figure 6-2 is fairly typical; living biomass rises to a peak of about 490-t/ha at 170 years after clear-cutting, drops to about 390-t/ha by 350 years, and then oscillates about a mean of about 370-t/ha.

## TOTAL BIOMASS ACCUMULATION

In Chapter 1 (Figure 1-10) we presented data on total ecosystem biomass (living plus dead) in the Reorganization and Aggradation Phases. We have no actual information on total organic matter for the later developmental Phases of Transition and Steady State. However, one line of reasoning suggests that total biomass, like living biomass, will exhibit steady-state properties. Living biomass represents the sole, direct and indirect, input of organic matter into the dead organic compartments (forest floor and dead wood). We have already indicated, above, that forest floor (Covington, 1976) and dead wood (A. Brush, personal communication) have a tendency to reach equilibrium. We also assumed that the mineral soil shows the same tendency. Therefore, we can expect that dead biomass as well as living biomass will fluctuate about some steady-state mean. Because of the oscillations of living biomass and the inherent lag in decay of organic matter, oscillations of total biomass will not be synchronous with those of living biomass. However, over the longer term, total biomass accumulation will be close to zero (Figure 6-1), and over time gross primary production will approximately equal ecosystem respiration.

We shall now examine the mechanisms by which such a steady state could arise and maintain itself through time. Although developed independently, these mechanisms turn out to be remarkably similar to those proposed by Alex Watt (1947) more than 30 years ago in his monumental paper, "*Pattern and process in the plant community.*"

## THE PLOT AS A UNIT OF STUDY

Thus far, we have discussed the northern hardwood ecosystem as an integrated unit equal to or greater than some minimum area necessary to maintain the integrity of the ecosystem as a functioning unit. For example, with JABOWA, which computes growth responses for 10 × 10-m plots, we simulated a curve of living biomass accumulation (Figure 6-2) for an area of 2-ha, or the sum of 200 plots. (Most of the biogeochemical data that we relate to the biomass accumulation model are based on watersheds that range from 12 to 40-ha.) We consider the living biomass accumulation curve in Figure 6-2 to represent an integrated response for the whole ecosystem. A quite different living biomass curve emerges if we follow an individual plot through time.

We must digress here to point out that nature does not work on the basis of JABOWA-sized plots but on the basis of irregularly-sized, naturally-occurring patches that result from disturbance. Thus, the forest that develops after clear-cutting might be considered one large even-aged patch resulting from exogenous disturbance. With time individual trees or groups of trees fall down, creating smaller patches of younger vegetation. Eventually, the whole area of the forest will be composed of patches of varying size and age. Although the discussion that follows is based on $10 \times 10$-m plots as defined by the JABOWA Forest Growth Simulator, we think it provides a logical approximation of the natural situation.

An individual JABOWA plot will *never* exhibit a prolonged steady state in which living or total biomass accumulation will approximate zero. That is, assuming no outside disturbance, it is highly probable that an individual plot would ultimately be dominated by a large old tree. Eventually, the old dominant dies and falls over. When it does, it creates a local disturbance, deposits a heavy concentration of dead biomass in a small area, and promotes an increase in resource availability somewhat like that characteristic for the ecosystem as a whole during the early part of the Reorganization Phase (Chapter 3). These changes are followed by rapid revegetation by numerous individuals, each with relatively small living biomass. With time, the plot undergoes thinning, living biomass accumulation occurs as the vegetation develops, and the biogeochemistry of the plot returns to conditions similar to those found during the Aggradation Phase (Chapter 2). Eventually, the plot again comes to be dominated by one or two massive trees, and the cycle repeats itself. It is

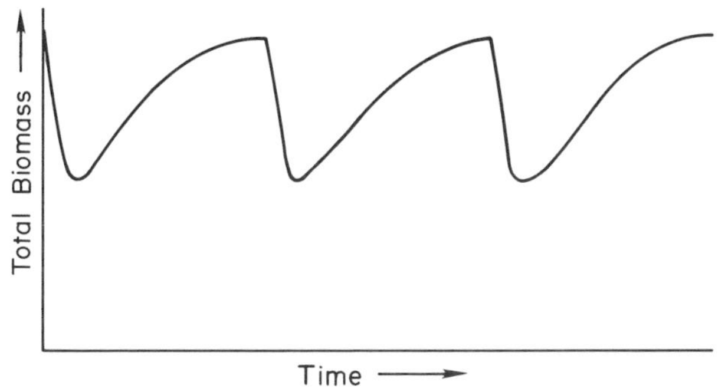


**Figure 6-3.** Hypothetical fluctuations in total biomass of an idealized patch. Assuming constant conditions of growth, death, and area, total biomass of an idealized patch shows cyclic fluctuations marked by a relatively steep loss in biomass (total respiration >GPP) and a more gradual rise in biomass (total respiration <GPP). The decline is initiated by death of a large old tree occupying the patch. These fluctuations are equivalent to Watt's upgrade and downgrade series (1947).

obvious that in an old, "undisturbed" ecosystem, every point within the ecosystem, whether it be included in a JABOWA plot or a naturally occurring patch, must continually, but irregularly, cycle between conditions approaching those of the Reorganization and Aggradation Phases (Figure 6-3).

## Even-Agedness: Synchronization of Development

Since clear-cutting imposes an even-aged condition on the ecosystem, the behavior of most plots is synchronized for about a century and a half after cutting. During the Aggradation Phase, as thinning continues and as the density of dominants declines, plots tend to be occupied by fewer and larger even-aged trees which date to the clear-cut. Thus, maximum living biomass for the entire developmental sequence is attained at the end of the Aggradation Phase, when a majority of plots support one or a few (more or less) even-aged trees of massive size. Forest structure at this stage might be considered as roughly equivalent to an old forest plantation where spacing has been adjusted to maximize living biomass accumulation. Interestingly, it seems that in the minds of many novelists, conservationists, foresters, and ecologists this type of massive, more or less even-aged successional forest is equated with "virgin," "climax," "pristine," or steady-state forest.

The end of the Aggradation Phase, when the ecosystem is dominated by old even-aged trees, must be considered a period of instability in the developmental sequence. The populations of even-aged dominants do not live forever; hence, the biomass peak is followed by a period of decline of living biomass (the Transition Phase), which results as even-aged dominants die out and are replaced by patches of young vegetation growing up from the forest floor. This leads to a drop in the living biomass of the ecosystem as the much lower weights of these patches are averaged into the whole.

Ecologists and foresters have reported that peak living biomass occurs toward the end of a pioneer stage (Loucks, 1970) or at the end of the life of an even-aged stand (Korstian, 1924). Apparently, the peak is reached at an age determined by the longevity of the species composing the stand, and after that point only a lower level of living biomass will be supported so long as the stand is free of exogenous disturbance [Korstian, 1924 (plots 11 and 12); Davis, 1966; Loucks, 1970; G. Furnival, personal communication].

The steady state is achieved after the progressive elimination of these old even-aged dominants and with the development of dominants of all ages and a more or less constant distribution of different-aged dominants on the plots that compose the ecosystem. Dead biomass, which is temporarily increased by accelerated input from the living biomass compartment (Figure 1-5) during the Transition Phase, eventually comes into equilibrium at a lower level. Thus total biomass in the Steady-State Phase is less than that at the end of the Aggradation Phase.

## Plot Status and Ecosystem Development

The developmental phase of the ecosystem is determined by the state of the naturally occurring patches comprising the ecosystem. To gain a more detailed view of the developmental process, we used JABOWA to simulate the behavior of living biomass in individual plots.

Based on the projected living biomass for the ecosystem as a whole at the end of the Reorganization Phase (Figure 6-2), we assigned plots with living biomass (dry wt) ranging 0 to 0.6-t/100-m$^2$ to State A. We use the term ***State*** to refer to the condition of a plot, in contrast to ***Phase***, which refers to the condition of the ecosystem as a whole. Plots in the State A condition may be dominated by herbs, shrubs, or young forest (e.g., one hundred 4-cm-dbh trees or three 18-cm-dbh trees = 0.6-t/100-m$^2$). Plots with a living biomass ranging from 0.6 to 4.9-t/100-m$^2$ are assigned to State B. These plots may be dominated by many small trees, a few intermediate-size trees, or a large tree (e.g., four 38-cm-dbh trees or one 67-cm-dbh tree = 4.9-t/100-m$^2$). Plots with a living biomass greater than 4.9-t/100-m$^2$ are considered to be in State C. State C plots may be dominated by one or two massive trees (e.g., two 69-cm-dbh trees or one 92-cm-dbh tree = 10.6-t/100-m$^2$). The plots in the various states are dynamically related to each other and have the developmental options shown in Figure 6-4.

The relative area (i.e., percentage of plots) in States A, B, and C and the trends of the States through time define the position of the ecosystem along the temporal scale of development (Figure 6-5). According to computer simulation, during the first 20 years the ecosystem is dominated by State A plots and is in the Reorganization Phase. For the next 150 years (Aggradation Phase), States B and C plots dominate. The abundance of B plots reaches a peak about 70 years after cutting, when they occupy 98% of the area of the ecosystem. Thereafter, their abundance declines sharply, while the abundance of C plots increases. By Year 170, C plots occupy about 55% of the ecosystem's area, and the ecosystem as a whole exhibits maximum biomass. State A plots temporarily disappear after Year 110.

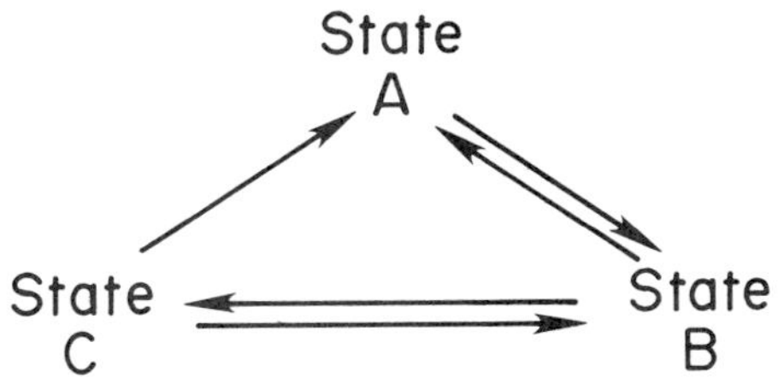


**Figure 6-4.** Developmental options for plots in various States of biomass accumulation. State A, 0–0.6 metric tons of living biomass per 100-m$^2$; State B, 0.6–4.9-t/100-m$^2$; State C, >4.9-t/100-m$^2$. All plots have the option of remaining in the same State with gain or loss of biomass as long as they stay within the limits of that State.

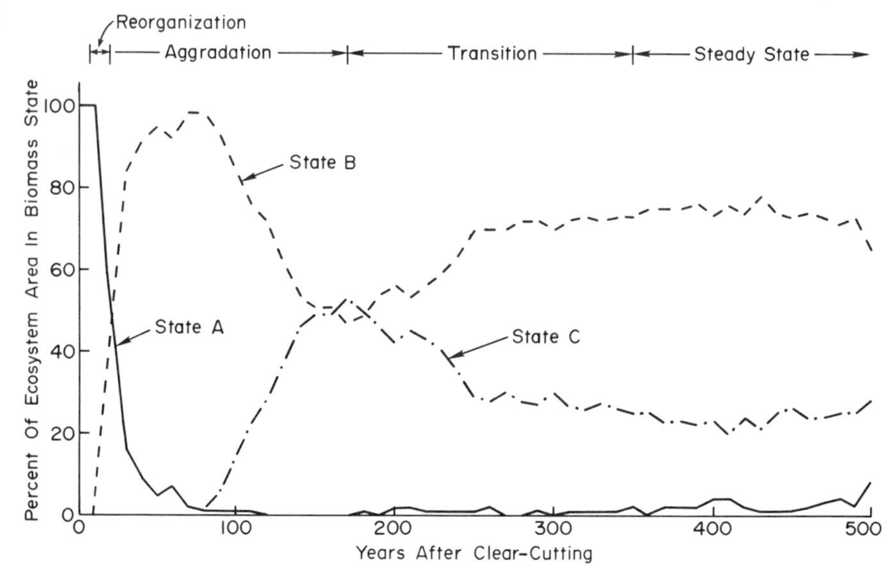


**Figure 6-5.** The percentage of ecosystem area in State A (<0.6-t/100-m²), State B (0.6–4.9-t/100-m²), and State C (>4.9-t/100-m²) in relation to time after clear-cutting. Based on JABOWA simulations involving two hundred 10 × 10-m plots.

This pattern of plot distribution during Reorganization and Aggradation Phases results because the majority of plots progressively accumulate biomass and tend to follow the developmental sequence State A → State B → State C (Figure 6-4), while relatively few plots lose biomass and drop to a lower state. This is largely a consequence of even-agedness, fairly long potential lifespans for some dominants, and natural thinning.

The Transition Phase shows a drop in the proportion of State C plots and a balancing increase in plots in States A and B. State A plots reappear after an absence of 60 years. The change in proportions of the three states during the Transition Phase is influenced by the death of many large individuals that were even-aged, quite old, and approaching the maximum age attainable by the species. In a sense, the Transition Phase may be thought of as a random opening of the forest due to age-induced senescence.

The developmental state of the individual plot, as reflected by its biomass accumulation, to some degree mirrors its biogeochemistry. Thus, a recently disturbed plot with little living biomass would tend to exhibit a biogeochemistry similar to that exhibited by the ecosystem as a whole during the early part of the Reorganization Phase, or a plot in State B would tend to exhibit a biogeochemistry similar to that of an ecosystem in the mid-Aggradation Phase. Thus, the proportions of plots in various states not only define the position of the ecosystem along the temporal scale of development but also strongly influence the biogeochemistry of the ecosystem.

However, plots in the same *state* (i.e., with about the same amount of living biomass) are not exactly equivalent biogeochemically when they are in different *phases* of development. To illustrate the point, we might consider a State A plot in a recently clear-cut ecosystem (Reorganization Phase) and in an older forest where several large trees have fallen and created a gap. In both cases, the State A plot would exhibit a temporary increase in resource availability after the disturbance; however, the nature and timing of the increase in resource availability would be different. The State A plot in the clear-cut would show fairly uniform conditions of water, nutrient, and energy availability throughout the area of the plot. The gap in the older forest, while exhibiting increased nutrient availability (Pigott, 1975), soil moisture (Douglass, 1967), and radiant energy at the ground level, would show a highly irregular pattern of availability relating to the geometry of the hole created by treefall. Penetration of radiant energy to the forest floor is dependent on the size, shape, and orientation of the opening in the canopy, while soil moisture and nutrient availability are influenced, among other things, by the distribution of functional roots remaining below the gap (Figure 6-6).

The patterns of decomposition of a State A plot in a clear-cut and in a gap in an older forest could be quite different. In a clear-cut, stems are removed, more solar energy reaches ground level (because all adjacent plots are clear-cut), and decomposition is rapidly accelerated in the forest

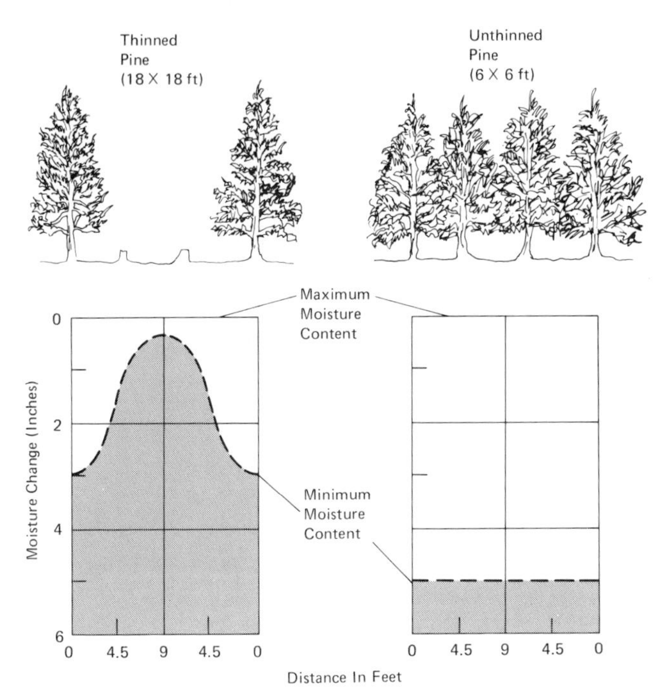


**Figure 6-6.** A comparison of soil moisture distribution beneath thinned and unthinned stands of loblolly pine. Thinning was done during April 1959 (from Douglass, 1967) 1 ft = 30.5-cm.

floor. Our estimates suggest that, for about 15 years, ecosystem respiration exceeds gross primary productivity and that after this time the plot begins to accumulate biomass (Chapter 5). Let us assume that a State A plot is created in the older forest by the fall of a large tree that uproots several smaller ones and that about 15 tons of dead wood are deposited in a gap of 300 $m^2$. This treefall represents an addition of 50,000 $g/m^2$ to the dead wood compartment, mostly in the form of slowly decaying trunks. If it took 30 years for this material to decay, there would be an annual loss of dead biomass of about 1700 $g/m^2$-yr for the plot, and to this should be added any net loss of organic matter in the soil. Even if we doubled the highest net living biomass accumulation measured in our 55-yr-old forest (435 $g/m^2$-yr; Table 1-1), it seems probable that the State A forest plot would require many more years than the State A plot in the clear-cut to show a net accumulation of total biomass. However, since nutrients were not removed from the forest plot, the subsequent rate of living biomass accumulation might be higher.

Although the proportions of plots or patches in various states of biomass accumulation are useful as a general guide to the biogeochemistry of an ecosystem, proportion alone may not be a precise indicator of biogeochemistry. Other factors (e.g., amount of dead wood, condition of surrounding plots) also influence a plot's biogeochemical behavior.

## The Shifting-Mosaic Steady State

In most discussions of steady state or "climax" ecosystems (e.g., Vitousek and Reiners, 1975), the condition of the steady-state ecosystem is loosely defined by one or two whole-ecosystem parameters, such as biomass accumulation equals zero or gross primary productivity equals total-ecosystem respiration. More often than not, authors are extremely vague or totally unconcerned about the fine structure of the steady-state ecosystem or the means by which the steady state might perpetuate itself through time. Yet such information is basic to any analysis of the energetic, biogeochemical, or stability relationships of the steady-state condition. Visualizing the steady state is not an easy task, and it is with some trepidation that we attempt an approximate characterization of what a hypothetical northern hardwood steady state might be like, and how it might perpetuate itself.

Based in part on a JABOWA simulation, we suggest that a steady state is achieved several hundred years after clear-cutting, when the standing crops of living, and total, biomass begin to oscillate about a mean (Figures 6-1 and 6-2). The proportion of the ecosystem in States A, B, and C remains more or less constant with time (Figure 6-5), but the state of any individual plot may change as it gains or loses biomass. For example, new plots in State A arise when gaps are created by the fall of large trees. An ecosystem in this dynamic, but relatively unchanging, condition we term the Shifting-Mosaic Steady State.

Several factors suggest that in a *real* steady-state ecosystem the proportion of total area in State A patches would be several times greater than that projected by the JABOWA simulation, which is based strictly on kill routines reflecting competition between trees (Botkin et al., 1972a,b). It does not include estimates of death or destruction arising from crushing effects as massive dominants fall (Figure 4-8). Earlier we reported observations where the fall of large dominants essentially cleared areas equivalent to three to five JABOWA-sized plots ($10 \times 10$ m). Also, using JABOWA we cannot consider local forces of disturbance that operate in any forest ecosystem—forces such as localized disease or insect attacks, lightning strikes, or attrition on trees surrounding gaps resulting from the weight of snow and ice and the pressure of high winds (Spurr and Barnes, 1973).

Given these considerations, the Shifting-Mosaic Steady State may be visualized as an array of irregular patches composed of vegetation of different ages. In some patches, particularly those where there has been a recent fall of a large tree, total respiration would exceed GPP, while in other patches the reverse would be true. For the ecosystem as a whole, the forces of aggradation and of decomposition would be approximately balanced, and gross primary production would about equal total ecosystem respiration. Over the long term, nutrients temporarily concentrated in small areas in fallen trunks would be made available to larger areas by root absorption and redistribution by litterfall.

The structure of the ecosystem would range from openings to all degrees of stratification, with dead trees concentrated on the forest floor in areas of recent disturbance. The forest stand would be considered all-aged and would contain a representation of most species, including some early-successional species, on a continuing basis.

The picture that the Shifting-Mosaic Steady State projects to us is remarkably like the "virgin" forest described in northern New Hampshire at the turn of the century:

> ... seedlings of tolerant species continue to exist under dense shade, but their growth is scarcely perceptible. When openings are made in the forest by the decay and death of the veteran trees, these suppressed seedlings begin to grow more rapidly under the stimulus of increased light. Reproduction of the more intolerant or light demanding species also takes place in the natural openings formed and these make up for lack of tolerance in rapidity of growth. Vacant places in the forest are thus soon filled up with a flourishing succeeding growth of tolerant and intolerant species struggling against each other for a place in the new stand (Chittenden, 1905).

## Corroborative Studies of Ecosystem Development

Studies of other forest ecosystems have yielded developmental patterns that conform reasonably well with the one we propose for the northern hardwood ecosystem (Figure 6-1). Watt (1925, 1947), after intensive study of development and structure of British beechwoods, constructed a diagrammatic model (Figure 10 in Watt, 1947) of succession in initially even-aged beech forest. Our Shifting-Mosaic Steady State seems to be almost a replica of his proposed old-age beechwood. R. B. Davis' study (1966) of stand dynamics in red spruce forests of coastal Maine provides a number of interesting comparisons with the total biomass accumulation pattern shown in Figure 6-1. He noted a trend from even- to uneven-aged conditions with an attendant decrease in living biomass. Davis suggested that the steady-state condition (his State C) is achieved at roughly 150 years but occurs only infrequently due to the prevalence of cutting, fire, and blowdown. Cooper (1913), Stearns (1949), and Goodlett (1954) proposed mechanisms similar to the Shifting-Mosaic Steady State for the perpetuation of boreal and northern hardwood forests.

The development of high-elevation fir forests in northeastern mountains (Sprugel, 1976; Sprugel and Bormann, 1979) shows some remarkable similarities to the Shifting-Mosaic Steady State in northern hardwoods. These diminutive fir forests occur near ridge lines and are subjected to strong winds; however, stands are not often flattened. Rather, under the influence of wind, waves of death and regeneration move through the forest at a fairly constant rate. New waves are regularly established downwind. Thus, every point in the ecosystem cycles through an upgrade and downgrade series just like every point in the Shifting-Mosaic Steady State. In terms of vegetation development, productivity, and biogeochemistry, the upgrade–downgrade series in the fir forest (Sprugel and Bormann, 1979) is analogous to that in the Shifting-Mosaic Steady State northern hardwood forests. The major difference is that the wind, which is generally unable to flatten the whole fir forest, organizes patches in the first state of development into long lines or waves that move through the forest at a fairly constant rate. The first state, equivalent to our State A, is followed by successive states, equivalent to States B and C, and then the first state is reestablished under the destructive influence of wind. In this way, every point in the fir ecosystem is turned over about once every 60 years. An area large enough to contain a complete set of developmental waves would exhibit steady-state characteristics, with GPP $\simeq$ total respiration. Thus, the high-altitude fir forest and the northern hardwood forest are similar in that the steady state is composed of patches in all states of development in more-or-less constant proportions. These forests differ in that patches in the northern hardwood ecosystem are randomly distributed and have a long cycling time (up to 300 years), while those in the fir forest are organized largely in waves and have a much shorter cycling time (about 60 years).

## TRENDS ASSOCIATED WITH ECOSYSTEM DEVELOPMENT

Eugene Odum (1969, 1971) presents a comprehensive, generalized model of ecosystem development in which autogenic development is characterized by an asymptotic curve of biomass accumulation and trends in a number of ecosystem attributes are associated with early and "mature" developmental stages. Odum's model and associated hypotheses have provided a healthy stimulus to the field of ecosystem ecology by organizing thinking on ecosystem dynamics and by providing a focus for further research. We now examine some conclusions drawn from the application of Odum's model to development of the northern hardwood ecosystem and make a comparison of these conclusions and those drawn from our model of biomass accumulation. We recognize that this comparison is not entirely parallel, since Odum's hypothesis is meant to apply to both primary and secondary succession and developmental and mature stages are not sharply defined. On the other hand, our model represents secondary development of a forested ecosystem after a catastrophic disturbance and involves fairly specific definitions of both structure and function. Nevertheless, some fundamental differences in interpretation of ecosystem behavior become apparent when we compare the conclusions drawn from these two models. Not all of these differences are limited to the specific case in question (i.e., northern hardwood ecosystems): some interpretations seem to have more general applicability.

### Energetics

*P/R Ratio.* The asymptotic model indicates that the P/R ratio (the ratio of gross primary production to total ecosystem respiration) is $>1$ in early stages and approaches 1 as development proceeds to mature stages. That model suggests that the P/R ratio should be an excellent functional index of the relative maturity of the system. Our model indicates that the P/R ratio does not follow a simple curve from $>1$ to 1. Prior to the Shifting-Mosaic Steady State, where $P/R \simeq 1$, there are two periods where the P/R ratio is $<1$ (i.e., the Reorganization and Transition Phases). For forest ecosystems in general, with large accumulations of living biomass, it seems probable that after any catastrophic disturbance which shifts large amounts of living to dead biomass, there will be a period of time when the P/R ratio will be $<1$. Consequently, the unqualified use of the P/R ratio as a simple functional index of development has strong limitations.

*Maximum Biomass.* The asymptotic model projects maximum biomass during the steady state, while the Hubbard Brook Biomass Accumulation Model predicts peak biomass at the end of the Aggradation Phase (Figure 6-1).

We think the idea of peak biomass prior to the steady state may have wide applicability to terrestrial ecosystems, and maximum attainable biomass accumulation may be the result of synchrony imposed upon the system by either nature or man. In northern hardwoods, peak biomass is achieved because even-agedness after clear-cutting imposes synchrony on all of the patches comprising the ecosystem. As a consequence, most patches, performing in lockstep, accumulate biomass for about a century and a half. Thereafter, ecosystem biomass drops as patches become more asynchronous and all-aged. A similar pattern of biomass behavior has been reported for British *Calluna* heathland after fire (Gimingham, 1972).

## Community Dynamics

*Habitat Diversity.* Odum (1971) broadly defines spatial heterogeneity and suggests that pattern becomes better organized with time. We may consider a plot in State A as equivalent to an open or brushy habitat, a plot in State B as analogous to a habitat dominated by vigorous trees of intermediate size, and a plot in State C as equivalent to a habitat dominated by one or two old trees. Using this definition of habitat, we suggest that, because of the even-agedness that follows clear-cutting, general habitat diversity remains low for a considerable period of time and then rises to a maximum during the steady state (Figure 6-7A).

Thus, even though the habitat of individual plots changes during the first century and a half after cutting, the overall habitat of the ecosystem at any one time tends to be fairly uniform (i.e., most plots are at about the same developmental state). Habitat diversity increases as the ecosystem becomes more and more a mixture of plots in all states of development. In effect, the attainment of the Shifting-Mosaic Steady State brings species that were temporarily separated in earlier developmental phases into close proximity. This would tend to diminish the effectiveness of temporal isolation along a successional gradient as an isolating mechanism in the evolution of species as suggested by Loucks (1970).

*Species Richness.* Odum's model using various diversity indices suggests that diversity will increase as development proceeds. Odum (1971, p. 148) also suggests that species diversity tends to be low in physically controlled ecosystems and high in biologically controlled ecosystems.

Species richness is a simple but useful measure of diversity (Whittaker, 1975). However numerous difficulties are encountered in trying to compare the total species diversity for different ecosystems, not to mention the problems of interpretation relative to ecosystem development. Detailed lists of all species in an ecosystem are never available and genetic material from seeds, spores, etc., are not normally included, so diversity is usually calculated on the basis of a few dominant or obvious groups. Similarly, we do not have information on the species richness of animals and microbes in the various stages of development for the northern hardwood ecosystem. Using species richness of rooted vascular plants (number of species occurring on 1-$m^2$ plots) as a simple index of diversity, we find the average number of species for randomly chosen 1-$m^2$ plots is lowest in the mid-Aggradation Phase (Figure 4-3), highest in recently clear-cut areas, and probably would be intermediate in the Shifting-Mosaic Steady-State Phase as the plots composing the ecosystem become all-aged (Figure 6-7B). The mixture of shade-tolerant, shade-intolerant, and intermediate species is a contributory factor to the high diversity of the recently-cut areas. In contrast, species richness of plants is lower during the mid-Aggradation Phase since shade-intolerant and many intermediate species are generally excluded from the ecosystem.

Questions of scale (size of landscape unit) and expression of potential diversity are immediately obvious. Clearly, a tree can fall during the developmental sequence and make an opening in the canopy. Reappearance of intolerant and intermediate species in such an opening might raise the species richness of the entire ecosystem, of which that patch was a

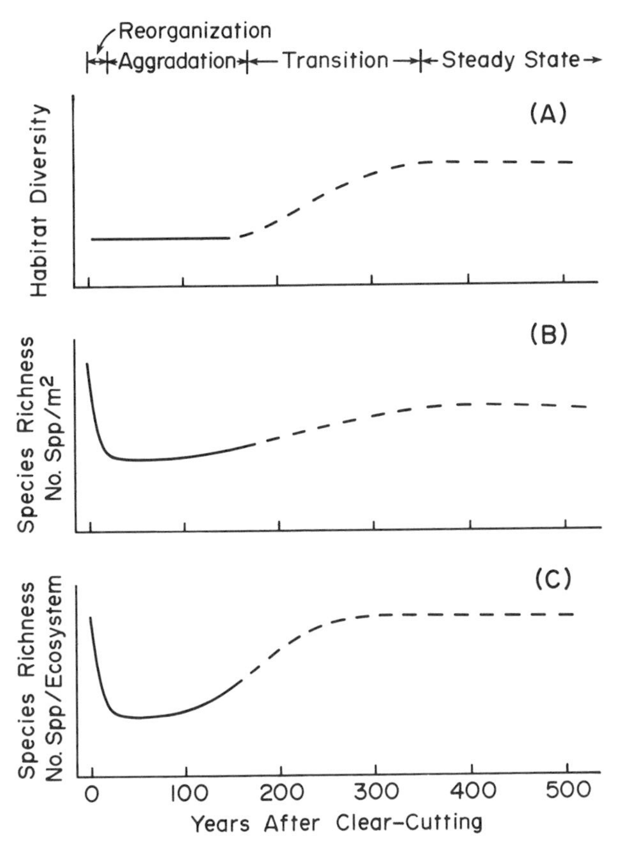


**Figure 6-7.** A schematic view of long-term trends of habitat diversity and species richness shown in relation to developmental phases after clear-cutting. Solid lines represent behavior patterns formulated from observation of extant stands, dashed lines show expected behavior of hypothetical stands.

component, to the highest level. Therefore, using the entire watershed-ecosystem as the reference unit, species richness would tend to be highest in both the recently-cut areas and in the Shifting-Mosaic Steady State (Figure 6-7C).

Considering only the actively growing plants, the observed patterns of species richness on either the ecosystem scale or the smaller scale provide an interesting contrast to Odum's model of ecosystem development (Figure 6-7C). Species richness for the northern hardwood forest tends to be lowest during the mid-Aggradation Phase of ecosystem development, when biotic control of metabolism and biogeochemistry is maximal (Chapter 2). Species richness peaks during the earlier, presumably more physically controlled portion of ecosystem development when biotic control is weakest. Obviously, there are many aspects of the relationships between species diversity and ecosystem development in need of refinement.

*Selection Pressure.* Odum (1969) suggests that during ecosystem development from early successional to mature stages natural selection pressure at first favors species with high rates of growth and reproduction ("r" selection). During more mature stages of ecosystem development, selection favors species with lower growth potential but better capabilities for competitive survival ("K" selection).

Our model suggests that natural selection during the Reorganization Phase might favor "r" selection, while "K" selection would tend to be more dominant in the Aggradation Phase rather than the Shifting-Mosaic Steady-State Phase. This is because early Reorganization is characterized by disturbed conditions and maximum resource availability while the Aggradation Phase is characterized by full occupancy of the site and strong competition. The Shifting-Mosaic Steady State represents a spatial mixture of the conditions that characterize the Reorganization and Aggradation Phases.

The exploitive and conservative plant-growth strategies discussed in Chapter 4 are roughly equivalent to "r" and "K" strategies. The developmental sequence after clear-cutting would at first favor selection for exploitive strategists (Table 4-8), which are able to capitalize on resources that have been temporarily increased (Chapter 5). Thereafter, because of increasing competition for water, light, and nutrients, selection pressure would favor the conservative strategists. In the Shifting-Mosaic Steady State, the selection pressure in patches, disturbed by treefall, would tend to favor the exploitive strategists, while throughout much of the remaining area selection for conservative strategies would prevail.

## Biogeochemistry

The Odum (1969) model does not discuss changes in hydrology or erosion during ecosystem development.

*Hydrology*. Water plays a major role in ecosystem dynamics as a requirement of plants, a vehicle of intrasystem nutrient cycling, and a vehicle of transport (export) of particulate matter and dissolved substances. We suggest that *major* variations in hydrologic output would occur only during the Reorganization Phase (Figure 6-8A). Oscillations in hydrologic export also may occur during the Transition and Steady-State Phases owing to local disturbance and regrowth resulting from treefall. Runoff from forest cutting may be directly related to the basal area

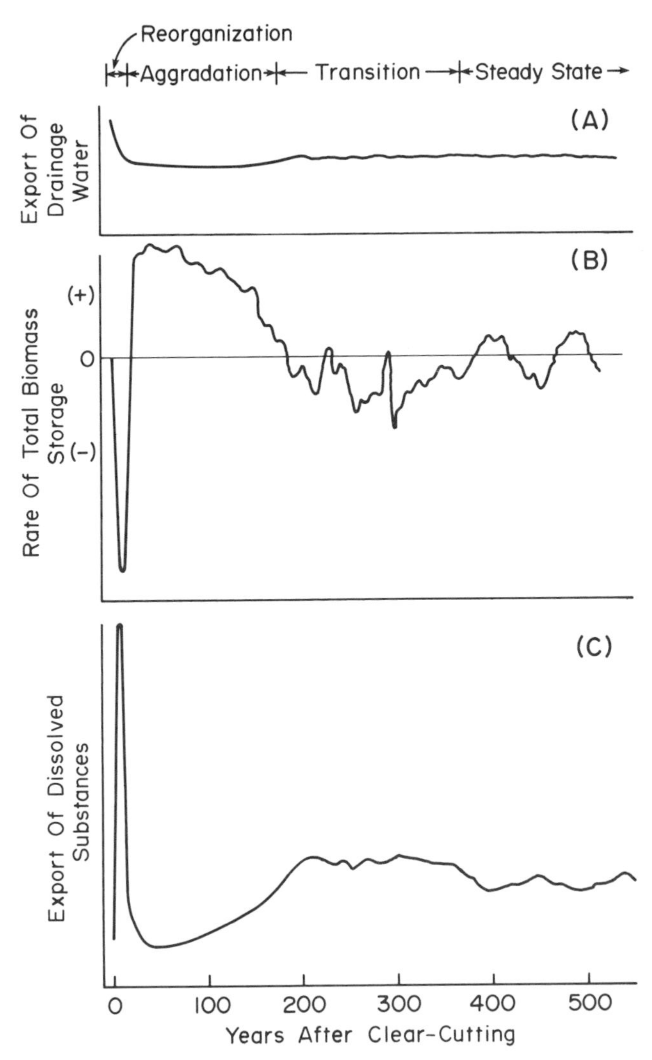


**Figure 6-8.** A schematic view of changes during 500 years after clear-cutting of a 60-yr-old northern hardwood forest: (A) export of drainage water; (B) rate of total biomass storage, with the ecosystem (+) accumulating biomass and (−) losing biomass; (C) the export of dissolved substances in drainage water.

removed (Hibbert, 1967), but Douglass (1967) reports that changes in runoff from forest cuts smaller than 20% of a stand's basal area are often difficult to detect at a gauging weir.

*Erosion.* Our data (Chapters 2, 3, and 5) indicate that the northern hardwood ecosystem is relatively resistant to erosion as long as disturbances do not seriously disrupt the forest floor. Even after clear-cutting (when it is done in a careful way), there is relatively little increase in erosion. In fact, we proposed that homeostatic responses after clear-cutting prevent greatly accelerated erosion from occurring. Erosion levels would rise modestly during the initial part of the Reorganization Phase but soon would be brought under control by developing vegetation. During the proposed course of ecosystem development, it seems probable that accelerated erosion would not occur, although isolated treefall (Chapter 4), particularly in the Transition Phase, could cause substantial local soil disturbance and promote local soil erosion.

*Weathering and Other Nutrient Sources.* This topic is not considered in detail in the Odum model. Weathering of minerals within the ecosystem is a primary source of available nutrients for elements with a sedimentary cycle. For some elements, precipitation also is a major source (Chapter 2). Biological fixation is a source of nitrogen; absorption and impaction by foliage is a source of airborne sulfur and other elements.

During the developmental sequence, it seems unlikely that all of these processes would proceed at the same rate. For example, during the initial part of the Reorganization Phase, the soil solution is rich in dissolved substances, while in the Aggradation Phase concentrations of dissolved substances are several-fold lower. These dissolved substances also are end products of weathering reactions and may affect weathering rates by mass action. Similarly, other processes may fluctuate with time. For example, the volume and architecture of the developing forest canopy may alter the rate at which aerosols are collected and thus deposited within the ecosystem.

The very low rate of removal of inorganic particulate matter from the ecosystem raises an interesting question about long-term weathering relationships. The ecosystem is being relatively enriched with Fe, Al, Si, and K (Chapter 2). Enrichment results in the development of a mineral soil profile in the form of $SiO_2$ and Fe and Al oxides. These relatively inert compounds accumulate near the top of the soil profile (i.e., the B-horizon), while in the same zone the amount of weathering substrate declines. Without introduction of new weathering substrate, the generation of available cations like Ca and Mg by weathering might tend to decline with time (development) and perhaps limit production and lower the amount of biomass the ecosystem can sustain.

How can new weathering substrate be brought to the surface? Several mechanisms operating within the ecosystem can accomplish this, for example, burrowing animals (not very important at Hubbard Brook) and treefalls that produce throw mounds, which in effect invert the soil profile and move some C-horizon material to the surface. These mechanisms, however, must ultimately be supplemented by erosion and transportation, which strip off, in the form of particulate matter, the upper layers (i.e., $SiO_2$, and Fe and Al oxides) and, in effect, lower the ecosystem in place thereby bringing more weathering substrate into the weathering zone.

A basic question arises: Is the rate of erosion sufficient to bring in adequate amounts of weathering substrate on a continuing basis? From a theoretical point of view, this is an exceedingly interesting question, since if the answer were no, this would imply that weathering substrate would become with time an increasingly important limiting factor. Perhaps, over millenia of time, periods of very intense erosion would be required to renew the weathering substrate and in effect renew the ecosystem.

*Export of Dissolved Substances and Particulate Matter.* The Odum model suggests that an important trend in ecosystem development is the "tightening" of biogeochemical cycles. Mature systems, as compared to developing ones, are proposed to have the capacity to entrap and hold nutrients.

We suggest that the trend toward tighter nutrient cycles does not parallel ecosystem development and that, instead, the greatest regulation over nutrient export occurs in the Aggradation Phase. Increased nutrient loss occurs during the Reorganization and Transition Phases; with biomass stabilization in the Shifting-Mosaic Steady State, export levels may be somewhat higher than in the Aggradation Phase. This pattern reflects in part the net rate at which nutrients are bound within the ecosystem in accumulating biomass (Bormann and Likens, 1967; Likens and Bormann, 1972a,b; Vitousek and Reiners, 1975; Vitousek, 1977).

There is little doubt that total biomass storage plays a major role in regulating nutrient export from the northern hardwood ecosystem. This relationship seems quite clear for young-successional ecosystems, and in earlier chapters we documented the correlation between decline in biomass and high nutrient export in stream water draining recently clear-cut ecosystems and the relatively low nutrient export from young forests which are rapidly accumulating total biomass.

However, more fundamental questions can be raised about the presumption that export patterns of terrestrial ecosystems are primarily a function of biomass storage. Other factors such as weathering rate, denitrification, or nitrification in addition to biomass storage may change with time. Thus, not only may biomass storage fluctuate as an ecosystem develops but so too may nutrient input and release. Streamflow and nutrient concentrations in drainage water also may fluctuate.

The export of dissolved nutrients during both the Reorganization and Aggradation Phases is reasonably well known by direct measurement (Chapters 2, 3, and 5), but almost nothing is known of longer-term phases. Nevertheless, we can make some approximations using the premise set forth that export of a particular element is equivalent to sources of new nutrients minus storage of nutrients within the ecosystem. This notion can be formalized:

$$\text{Export} = \text{Sources} - \text{Storage}.$$

Nutrient storage (exclusive of nutrients stored in rock and secondary minerals) in the northern hardwood ecosystem is largely accounted for by nutrients bound in living and dead organic matter and held on exchange sites, mostly organic in nature (Hoyle, 1973). Proposed biomass storage relationships for the ecosystem through time are shown in Figure 6-8B.

Assuming that sources of net nutrients are *constant*, as did Vitousek and Reiners (1975), for terrestrial ecosystems in general, the following pattern of nutrient export shown in Figure 6-8C emerges. This pattern, however, differs from that presented by Vitousek and Reiners, since they used an asymptotic model of biomass accumulation, while we used the Hubbard Brook Biomass Accumulation Model and attempt to take into account changes in biogeochemistry and hydrology. The Reorganization Phase for northern hardwood forests is characterized by a high rate of total biomass loss over a short period of time and by a maximum rate of export of dissolved substances and particulate matter in drainage water. A net loss of nutrients from the ecosystem results. The Aggradation Phase exhibits a high rate of total biomass storage throughout most of the period but tapers to very little at the end of the Phase. Coincident with this, the ecosystem accumulates nutrients, and the export of dissolved substances in drainage water during the early part of the Aggradation Phase is at a minimum for the entire developmental sequence. However, dissolved substance export may increase toward the end of the Phase as biomass storage decreases. The Transition Phase is marked by a net loss in total biomass over a fairly long period of time. Based on JABOWA simulations, this loss of biomass can occur somewhat abruptly or fairly regularly throughout this Phase. These options make it difficult to propose a detailed model of drainage-water chemistry for this period, but for the period as a whole the ecosystem probably loses substantial amounts of nutrients that were stored in biomass (Table 6-1), and the export of dissolved substances in drainage water may be higher than that of the Aggradation Phase. The Steady-State Phase is characterized by both storage and loss of total biomass, cycling about the steady-state mean. Over the long term, biomass storage would remain more or less constant, and the standing crop of nutrients held therein would do the same. On this basis, nutrient export in stream water would be generally greater than that of the Aggradation Phase but would fluctuate according to the cyclic behavior of biomass storage and rates of endogenous disturbance.

Treefall during the Transition and Shifting-Mosaic Steady-State Phases could produce appreciable disturbance of the forest floor; however, because of the isolation of these disturbed areas from each other it seems improbable that individual treefalls would result in greatly accelerated removal of eroded material from the ecosystem.

**Table 6-1.** Estimated Annual Net Loss of Nutrients Due to Net Loss of Biomass During the Transition Phase, Year 170 to Year 350[a]

| | Nutrient Content in Living Biomass (kg/ha) | | |
|---|---|---|---|
| Nutrient | 170-yr-old Ecosystem | 350-yr-old Ecosystem | Estimated Annual Net Loss (kg/ha-yr) |
| N | 1540 | 1123 | 2.31 |
| Ca | 1447 | 1146 | 1.67 |
| K | 548 | 473 | 0.42 |
| S | 182 | 142 | 0.22 |
| P | 163 | 130 | 0.18 |
| Mg | 142 | 118 | 0.13 |
| Dry biomass ($\times 10^3$) | 497 | 383 | 0.63 |

[a] *The calculation assumes dead biomass pools at Years 170 and 350 are in balance. Estimates were made using JABOWA, the forest growth simulator.*

Because of the relationships between biomass storage and nutrient export in stream water it has been suggested that stream-water concentrations can serve as an index of the developmental state of the ecosystem (Vitousek and Reiners, 1975; Vitousek, 1977). Although stream-water concentrations seem to reflect the developmental state of relatively young, even-aged northern hardwood stands, this parameter has limited applicability to older stands. For example, using Vitousek and Reiners' assumptions, the same level of nutrient concentrations in stream water could occur at various points in our biomass accumulation curve, i.e., portions of the curve having the same rate of biomass accretion or loss (see Figures 6-1B and 6-8B). Similar rates of accretion and loss can occur in various developmental phases. Thus, for interpretive purposes, data on stream-water concentrations must be accompanied by quantitative data on the ecosystem and its structure and development. Necessary data involve ecosystem (watershed) area, the proportion of the ecosystem in patches at different stages of development, relative age of patches, and so forth.

Moreover, relatively small changes in the export of nutrients from ecosystems are influenced by a number of factors. These include temporal changes in water chemistry not related to ecosystem development. For example, short-term changes in microclimate, such as soil frost, may appreciably change stream-water chemistry and obscure long-term trends (Likens et al. 1977). Changes in stream-water chemistry related to ecosystem area above the sampling point also must be considered. Ecosystems of markedly different size may yield appreciably different stream-water concentrations (N. Johnson, personal communication), although export per unit area may be similar (J. Sloane, personal communication). Without a detailed analysis of the overall ecosystem, then, it seems to us unwise to label an ecosystem "old-age," "steady state," or "climax" and then use short-term, stream-water chemistry to characterize the proposed condition.

To sum up, nutrient patterns during ecosystem development as deduced from presumed biomass-storage patterns and measured stream-water chemistry leave much to be desired, but they may provide a rough estimate of long-term nutrient relationships. It is obvious that we need more intensive study of both biotic and abiotic functions and their interactions in ecosystems before we can fully appreciate the dynamics of nutrient cycling and its change with time.

## Stability

Every ecosystem is subject to an array of external energy inputs—radiant energy, wind, water, and gravity. All of these represent potentially destabilizing forces that may destroy or diminish ecosystem organization or reduce its substance. For an ecosystem to grow or to maintain itself, it must be able to channel or meet these potentially destabilizing forces in such a way that their destructive potential is not realized. Control or management of destabilizing forces is the essence of ecosystem development and stability. Stated another way, stability is a function of the capacity of the ecosystem to control or channel the forces impinging on it.

We use relative export characteristics as indicators of system stability, with low and fairly constant export of dissolved substances and particulate matter signifying a stable system and higher and irregular export characterizing a less stable system.

The Reorganization Phase (Chapter 3) is the "loosest" of the phases. It is characterized by (1) the highest concentrations of dissolved substances in drainage water, (2) the lowest levels of transpiration, and the highest levels of streamflow and erosion, and (3) rapidly changing biogeochemistry as revegetation proceeds. The biogeochemistry of the Reorganization Phase reflects the complete destruction of the canopy by clear-cutting, a decrease in total biomass as decomposition outweighs gross primary productivity, and a net loss of nutrients. "Looseness" of the system is associated with high species richness and dominance by plants with exploitive strategies.

The Aggradation Phase, particularly the early part, is characterized (Chapter 2) by (1) low, highly predictable and fairly constant concentrations of dissolved substances in stream water draining the ecosystem, (2) the highest levels of transpiration and the lowest levels of streamflow, (3) low and predictable net export of dissolved nutrients, and (4) very low levels of erosion (i.e., particulate matter export). All of these criteria indicate that the early aggrading ecosystem is a "tight" ecosystem. These biogeochemical characteristics are strongly influenced by a more or less continuous forest canopy, a high rate of biomass accumulation, and a high rate of net nutrient accumulation in living and dead biomass. It is interesting to note that the "tightness" of the ecosystem, or stability by our definition, is associated with dominance by species employing the conservative strategy and that species richness is at a minimum.

Based on projected biomass behavior alone (Figure 6-8B), Transition and Shifting-Mosaic Steady State would occupy positions somewhere between Reorganization and Aggradation, with export characteristics falling somewhere between the two extremes (Figure 6-8C).

The Aggradation Phase exhibits the lowest biogeochemical exports (and by our definition has the highest degree of biogeochemical stability), but, in a sense, this phase may be thought of as culminating in a biologically unstable condition. That is, the even-aged aggrading forest can survive a long time and during that time exhibit maximum biogeochemical regulation, but with time the ecosystem is populated with many now old, more or less even-aged trees, each with a substantial destructive potential (Chapter 4). Such a situation has many aspects in common with a population of animals just prior to a population crash, with the exception that the mixture of the species comprising the old aggrading northern hardwood forest would tend to spread such a crash over many decades. The Transition Phase then might be thought of as a prolonged crash period leading to an all-aged steady state with less regulation over biogeochemical export.

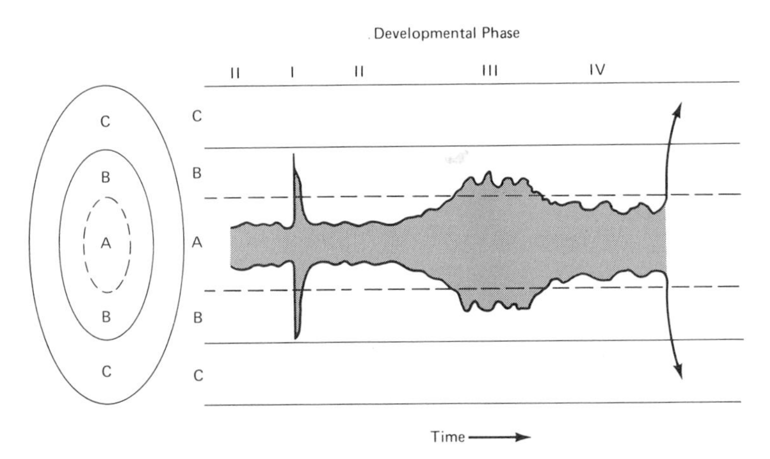


**Figure 6-9.** *Left*: A graphic model of stability–resilience after Margalef (1969) and Hill (1975). The meaning of areas A, B, and C is discussed in the text. *Right*: The graphic model on the left is extended through time as a cylinder. The diagram is a bisect of the cylinder showing the position of the developing northern hardwood ecosystem within areas A, B, and C. The bisect represents a second-growth forest that is clear-cut and allowed to develop to the Shifting-Mosaic Steady State without external interference. The arrows to area C represent the system's response to a severe perturbation (e.g., a hot fire followed by severe erosion; see Figure 6-10).

(A)

Grantham Mountain
806 m
Burned Area (1953)
1977 Photograph
from Ground
1977 Aerial
Photograph

(B)

Grantham Mountain

(C)

**Figure 6-10.** Grantham Mountain, Plainfield, New Hampshire. A fire burned over the summit of this mountain in 1953. (A) The extent of the fire is shown as the black region in an infrared photo taken at the time of the fire. (B and C) Photographs taken in the autumn of 1977 showing the burned area near the summit, much of which is now bare rock. Aerial photos taken in 1941 and a forest-type map made in 1947 show few bare or open spots in the area before it was burned in 1953 (A. Plumb, personal communication). The burned area previously supported modest spruce and spruce–hardwood forests on thin soils. The fire and subsequent erosion drove portions of the ecosystem over the resiliency threshold (Figure 6-9) and converted the forest ecosystem into a bare rock–shrub system. Centuries will be required for these bare-rock areas to once again support forests. (Photos by Adrian Bouchard.)

Treefall, which is essential to the maintenance of the Shifting-Mosaic Steady State, also may play an important role in long-term biogeochemical relationships. For example, as mentioned previously, with development, weathering substrate may tend to be isolated from the most active zone in the soil. Through treefall, however, this substrate may be brought to the surface in throw mounds and made more available to the functioning ecosystem.

One definition of stability is concerned with the ability of the ecosystem to return to its original condition after an external perturbation. This is usually referred to as stability–resilience. Resilience is concerned with the ability of the system to absorb disturbance and still persist within threshold boundaries. Threshold limits are discernible only after perturbation of the ecosystem by severe external stress (Hill, 1975). Margalef (1969), to clarify the concept of resilience, proposed the idea of mapping the stability of an ecosystem into three areas. Hill (1975) summed Margalef's proposal in the following terms:

> Within the central area [A in Figure 6-9], the system may oscillate in a spontaneous manner as it adjusts to predictable environmental changes, or to cycles generated by interactions between species . . . The system moves, if subjected to moderate stress, to an area B, which surrounds area A. However, such changes can be assimilated or buffered, enabling the system to return to area A. If the stress is severe enough the system moves into area C, outside B, and the properties of the system may seek new steady state levels instead of returning to original conditions. Consequently, the boundary between B and C may be regarded as a threshold of system resilience to particular stress conditions.

To gain a temporal concept of stability–resilience relationships during secondary development, we have added a third dimension, time, to Margalef's two-dimensional model. The model may be conceived as a cylinder composed of three shells—A, B, and C. The length of the cylinder is the time dimension. A bisect of the cylinder yields ecosystem responses through time.

Using ecosystem export characteristics (Figure 6-8) as a black box measure of ecosystem response, we have mapped the behavior of a second-growth northern hardwood forest ecosystem that was clear-cut and allowed to develop to the Shifting-Mosaic Steady State but ultimately subjected to a disturbance that drove it beyond the resilience threshold of area C (Figure 6-9).

Clear-cutting evokes a sharp response in export behavior, but the response is within the threshold limit. The mechanisms discussed in Chapter 4 rapidly restore the ecosystem to area A. Even though there are changes in species composition, the even-agedness, more or less continuous canopy, and high rate of biomass accumulation keep the ecosystem near the core of area A. Gradually, however, with the development of an ecosystem composed of large even-aged trees and a reduction in the rate of biomass accumulation the system moves toward area B. Next, the relatively high death rate of large trees and the loss of biomass move the system into area B. Gradually, however, the system returns to area A as it becomes all-aged. However, it is less close to the core of area A than the Aggradation Phase. This is because it is not consistently accumulating biomass, and it is more irregular than the Aggradation Phase because it tends to cycle around the steady-state mean. At any time, the ecosystem might be driven over the resiliency threshold by a major disturbance such as a hot fire followed by extreme erosion (Figure 6-10). This would reduce the ecosystem to a primary condition and prevent its return to original conditions for a very long time if at all.

## Summary

1. A biomass accumulation model is presented showing the development of a steady state after clear-cutting in a northern hardwood forest. This model departs in several major aspects from the widely accepted asymptotic model of ecosystem development.
2. When the ecosystem is viewed as a collection of patches, the steady state is a condition in which the proportions of patches in different states of development remain more or less constant, yet with each patch continuously undergoing a sequence of biomass accumulation and loss. We call this condition of the ecosystem the Shifting-Mosaic Steady State.
3. Developmental trends predicted from the Hubbard Brook Biomass-Accumulation Model are compared with those predicted by the asymptotic model. There are a number of basic differences regarding energetics, community dynamics, and ecosystem export patterns.
   a. The maximum amount of biomass a northern hardwood ecosystem can support is not found in the steady state, as predicted by the asymptotic model, but in an earlier phase. Maximum supportable biomass is a product of synchronous behavior of patches composing the ecosystem, and this behavior might be common to many kinds of forest ecosystems.
   b. The use of the P/R ratio as a simple index of ecosystem development has strong limitations.
   c. In general, habitat diversity remains low throughout much of the developmental period and then rises to a maximum in the Steady-State Phase.
   d. Species richness of plants for the entire ecosystem is highest in recently-cut areas (Reorganization Phase) and the Shifting-Mosaic Steady State and lower in the mid-Aggradation Phase. Average number of species per randomly chosen plot (1 $m^2$) is highest during the Reorganization Phase, lowest during the mid-Aggradation Phase, and intermediate in The Steady-State Phase.
   e. In terms of export from the ecosystem, control during the early Aggradation Phase of development is "tighter" than in the Steady State. This is due to even-agedness, a more or less continuous canopy, and a high rate of biomass accumulation.
4. Although changes in biomass storage affect nutrient export from the ecosystem, a number of unknowns (e.g., rates of weathering, nitrogen fixation, denitrification, and so forth) complicate predictions of biogeochemical flux and cycling, particularly in later phases of ecosystem development.
5. Stability–resilience relationships of the Hubbard Brook Biomass Accumulation Model change through time because of endogenous and exogenous disturbances.

CHAPTER 7

# The Steady State as a Component of the Landscape

Models of ecosystem development usually portray autogenic succession as an orderly progression of biologic changes (e.g., Odum, 1969; Woodwell, 1974). The macroenvironment within which development occurs is presumed to be more or less constant throughout the autogenic sequence. Yet every terrestrial ecosystem is subjected to a range of disturbances varying from those that barely alter the structure, metabolism, or biogeochemistry of the ecosystem to those that wholly or dramatically change the system. Defining "disturbance" is itself a considerable problem, because it is difficult to draw a line between biological and physical-chemical events that may be considered within the scope of autogenic development and other events that might be considered to seriously deflect the autogenic pattern. In developing the Hubbard Brook Biomass Accumulation Model (Chapter 1) of ecosystem development, we followed the procedures of Odum (1969), Botkin et al. (1972a,b), and Woodwell (1974) and emphasized autogenic development, while deemphasizing exogenous disturbance. This was a necessary decision if our model was to reflect an uninterrupted sequence from the initiation of secondary development to the establishment of the steady state.

An imposing and growing body of literature, however, has emphasized the importance of fire and wind as common, widespread perturbations that historically have shaped the structure and function of North American forests. Several hypotheses, while recognizing the importance of autogenic development, have proposed that, historically, forest ecosystems have been destroyed and restarted at irregular, but relatively short, intervals by catastrophic disturbances. The reality and even the intellectual value of the steady-state concept have been questioned (Stephens, 1955; Raup, 1957; Loucks, 1970; Heinselman, 1973; Rowe and Scotter, 1973; Wright and Heinselman, 1973; Henry and Swan, 1974).

Loucks (1970) proposed that before settlement by Europeans forests in Wisconsin were repeatedly but irregularly burned at relatively short intervals (Figure 7-1). These catastrophic disturbances presumably truncated the development of the ecosystem, in effect rejuvenating it and restarting it at an earlier developmental state. Recovery patterns for production and biomass accumulation and some species and community parameters of the ecosystem are portrayed as wave phenomena, irregularly restarted by fire at relatively short intervals (i.e., a curve that rises to a peak, declines, and rises again, and so on). Loucks' hypothesis neither allows nor considers an autogenically derived steady state but seems to imply that it occurs only rarely and is composed primarily of tolerant plant species. Accurate fire histories compiled in the western region of the Lake States support the conclusion that a steady state was rarely achieved in presettlement forests (Frissell, 1973; Heinselman, 1973).

A similar, but less well-documented hypothesis, has arisen from studies in central New England, where strong cyclonic winds (e.g., hurricanes) are thought to interrupt autogenic development regularly prior to the steady state (Stephens, 1955, 1956; Raup, 1957; Henry and Swan, 1974).

These hypotheses raise fundamental questions about the relationship between autogenic development and the incidence of major disturbance and questions as to the utility of the steady-state concept in the interpretation and management of northern hardwood ecosystems. Is the steady state merely a theoretical concept with little or no reality in nature? Were historic northern hardwood ecosystems mostly characterized by wave-form patterns triggered by random disturbance at intervals much shorter than the time necessary to achieve the steady

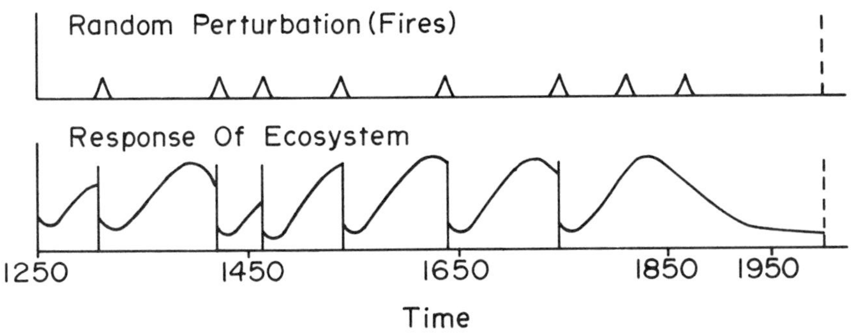


**Figure 7-1.** A model of transient phenomena and ecosystem responses (in biomass, primary productivity, diversity of seedling layer, etc.) as a function of time for southern Wisconsin forests. Ecosystem development is restarted by each perturbation, and the whole series of phenomena and responses is thought to constitute a stable system. With fire suppression by man, the cyclic process is disrupted (after Loucks, 1970).

state? Are recurring catastrophes promoted by autogenic development patterns that progressively immobilize nutrients or increase fire or wind susceptibility? Has European man substantially changed the historic pattern? To consider these difficult questions we must examine the historical record to determine, as best we can, the regularity with which pre-European northern hardwood forests in the White Mountain region were destroyed by catastrophic fires or winds.

Our evidence suggests that substantial areas of the region were free of catastrophic disturbances for long periods of time. This does not establish the widespread occurrence of steady-state ecosystems in pre-European times, but it does suggest that the potential for their occurrence existed. We also consider ways in which European man altered the forest. One of the major effects is the repeated imposition of even-agedness on the ecosystem by a variety of land-use practices. As a result, the biomass of the ecosystem tends to cycle through a sequence of peaks and troughs.

## EXOGENOUS DISTURBANCE DEFINED

Effects of disturbance are extremely complex since many factors, both biological and physical, may act to make trees more susceptible to the action of physical forces. These include biological processes like competition, senescence, insect and disease attack, and simple growth in size, as well as climatic fluctuations that are unfavorable to growth, such as hard frosts, severe drought, or prolonged periods of unfavorable weather. Physical disturbance itself can act as a positive feedback by making a tree more susceptible to subsequent disturbance by weakening its competitive position—by breaking its limbs and reducing its leaf area or by increasing its susceptibility to disease and insects.

Most agents of disturbance occur as a continuum; for example, winds may range from mild breezes to winds of hurricane force. Consequently, it is difficult to draw a line between what we call endogenous disturbance and exogenous disturbance. We have defined endogenous disturbance as disturbance caused by the fall of trees weakened or killed in the autogenic development process. The fall of these trees might be caused by their own weight or loading by wind, ice, or snow. Whatever the final cause that tips them over, their fall is inevitable even in the most protected of locations. Endogenous disturbance is included in our developmental model and is an integral part of the developmental process. Exogenous disturbance is considered as resulting from forces that may be thought of as external to the developmental process, such as intense winds or fires. These forces not only hasten the death or fall of trees weakened in the developmental process, but they damage or kill many trees that were in reasonably healthy condition prior to the event. These disturbances also affect, in a fairly intensive way, a variety of ecosystem processes. Exogenous disturbances of this kind are not included in our developmental model. We shall now consider intense winds and fire as factors that shape the northern hardwood forest ecosystem in the White Mountain region.

## COMPARATIVE EFFECTS OF MAJOR PERTURBATIONS

Before proceeding, it is useful to recognize that the major perturbations being considered here (e.g., fire, wind, or clear-cutting) may have appreciably different effects on subsequent behavior of the ecosystem. Although all bring about a temporary increase in the ratio of oxidation to primary production, this is achieved in very different ways. Fire may largely destroy aboveground biomass and reduce gross primary productivity to nearly zero. At the same time, fire removes nutrients by volatilization. Oxidation by fire is very rapid, and in some cases fire also might promote increased biological oxidation (decomposition) of organic matter remaining in the soil. The increase in resource availability (e.g., soluble ash products) tends to be rapid, but may be quickly flushed from the ecosystem. Clear-cutting also may destroy most aboveground vegetation, and substantial quantities of nutrients are removed in wood products. In this case, oxidation is by biological processes and occurs less rapidly than in fire but is still relatively fast. The forest floor is the major site for increased decomposition (see Chapter 3), and the increase in resource availability may be spread over a somewhat longer time than that occurring with fire.

Most intense winds are erratic in their effects (Figure 7-2). Many or all of the larger trees may be blown down, but some residual trees are often left standing (Spurr, 1956a; Spurr and Barnes, 1973). Immediate effects on GPP would seem to be less severe than with intense fires or clear-cutting. The remaining forest may be a patchwork of open areas and remaining vegetation with a tangle of leaning and fallen trees. Little material is removed from the ecosystem. Under these circumstances, it seems likely that the decomposition response of the forest floor would be more subdued than after clear-cutting. However, massive amounts of tree trunks may be added to the dead wood compartment. As discussed earlier, decomposition of this material may keep the oxidation/GPP ratio for the ecosystem as a whole $>1$ for decades after the event. In turn, this might imply a relatively small increase in resource availability spread over several decades (i.e., as the total biomass of the system declines, there is a net release of nutrients).

In terms of age structure, both fire and clear-cutting can result in essentially even-aged stands or stands with a few age classes (Gilbert and Jensen, 1958; Frissell, 1973; Heinselman, 1973; Wright and Heinselman, 1973). Wind also may bring about an even-aged stand, but in many

(A)

(B)

**Figure 7-2.** Wind-damaged northern hardwood forests in northern Pennsylvania showing the erratic nature of wind damage. (A) An area of about 1 ha blown down in a strong windstorm. Near the center all trees are flattened, but toward the periphery some intact trees are mixed with windfalls. (B) A wind-damaged site showing many smaller trees released by the destruction of the overstory (Photos courtesy of the Hammermill Paper Company.)

cases the ecosystem that develops after an intense wind event may be composed of an occasional overstory tree and released understory trees mixed with an even-aged group of seedlings, sprouts, and advance regeneration that becomes established or released in openings (Spurr and Barnes, 1973).

## DISTURBANCE IN PRESETTLEMENT NORTHERN HARDWOOD FORESTS

To gain a better focus, we shall examine the occurrence of catastrophic fire and wind in and around the northern hardwood forest region and then narrow our perspective to the hardwood forest of the White Mountains of northern New Hampshire.

### Fire

The northern hardwood forest grades into the boreal forest along its northern border. Within these areas, catastrophic fires, both natural and man-made, were and are common throughout the area (MacLean, 1960; Rowe and Scotter, 1973). Fire histories reconstructed by Heinselman (1973) and Swain (1973) indicate that fire has been an important component of transitional forests between boreal and deciduous zones for thousands of years. Heinselman's detailed study of the Boundary Waters Canoe Area in northwestern Minnesota indicates a natural fire rotation of 100 years in presettlement times; i.e., on the average, an area equivalent to the whole area was burned over every 100 years. Between 1681 and 1894, 83% of the area burned in nine fire periods. Collectively these data suggest that only a relatively small proportion of the transitional forest is free from fire for periods exceeding several hundred years. Heinselman (1973) questions the utility of the Clementsian concept of climax under these recurrent-fire conditions.

Old even-aged stands of white pine occur throughout the Great Lakes region and are generally thought to have originated after intense fires (Chapman, 1947). Based on even-aged stands Graham (1941) postulated a great fire in the upper peninsula of Michigan between 1400 and 1600, and Maissurow (1941) speculated that forest fires burned through 95% of the virgin forests of northern Wisconsin. Hough and Forbes (1943) reported an even-aged white pine stand in Tionesta Forest in northwestern Pennsylvania that originated from a fire in 1644.

In more detailed studies involving dating by fire scars and stand age, Frissell (1973) reports that fires occurred on the average every 10 years between 1650 and 1922 in Itasca Park in central Minnesota and occurred throughout the area of the Park. In common with Heinselman's findings, fires often resulted in the subsequent establishment of even-aged pine stands and in a presettlement mosaic of largely even-aged stands dating from different fires. Lutz (1930), based on a fire-scar study, reported 40 fires between 1687 and 1927 at Heart's Content Forest in northwestern Pennsylvania. Five of these fires, 1749, 1757, 1872, 1903, and 1911, were considered to be severe, and two were thought to be associated with severe drought years. Stearns (1949) in his study of old-aged northern hardwood forests in Wisconsin reported serious fires after lumbering but proposed that fire was overrated as an initial destructive agent in old-aged hardwood stands.

The record of fire occurrence in northeastern forests, based on even-aged stands, fire scars, or charcoal, is far less complete than for the western extension of the northern hardwood forest. Cline and Spurr (1942) suggest that more or less even-aged stands of old-growth white pine in southwestern New Hampshire originated from fires. More recently, Henry and Swan (1974), working in the same area and using tree rings and the occurrence of charcoal in the soil, dated one of these fires as occurring in 1665.

Based on early historical writings, it seems probable that fires started by Indians were common in presettlement times in southern and central New England and westward to Pennsylvania (Lutz, 1930; Day, 1953; Spurr, 1956b; Thompson and Smith, 1970). Kalm (1770) reported extensive Indian fires along the southern Champlain Valley of New York State and Vermont in 1750.

Historical evidence for the deliberate use of fire by the Indians in northern New England, however, seems to be lacking (Day, 1953; Brown, 1958a). Indian populations in the mountainous regions were small and migratory. In the vicinity of Hubbard Brook, Indians apparently spent the fall, winter, and spring months in the river valleys. Maize, beans, squash, and pumpkin were planted in early spring, and then most of the group migrated to the coast for the summer season, returning in the late summer (Likens, 1972b).

In river valleys surrounding Hubbard Brook, early explorers noted the occurrence of white pine stands of various size and age (Brown, 1958a). These may have originated after abandonment of Indian garden sites or after fire that had been used to clear land. Day (1953), however, suggested that the usual incentives for Indians to burn the forest were lacking in northern New England—agriculture was less practiced, summer travel was by canoe rather than overland, winter travel was by snowshoe and was not hindered by underbrush, and deer hunting took the form of stalking or still hunting rather than driving. Comments from land surveyors on the nature of the land along survey lines of township borders made from 1783 to 1787 make no mention of fire or fire scars in the presettlement forest of northern Vermont (Siccama, 1971).

The available literature suggests that fires were common in presettlement times throughout much of the northern hardwood and boreal region surrounding the area of the White Mountains and northern Vermont. Within the White Mountain region there is very little evidence, vegetational or historical, for the common occurrence of widespread fire in presettlement northern hardwood forests. Certainly, there are no fire histories comparable to those found by Heinselman (1973) and Frissell (1973) in Minnesota.

## Wind

Wind effects, because of their relative independence from control by man, will be considered in terms of both pre- and postsettlement time periods.

Stearns (1949), in a study of land-survey records, reported considerable wind damage to presettlement northern hardwood forests in northeastern Wisconsin. Surveyors' maps made about 1859 show areas of blowdown covering one to several square kilometers, and their notes record many instances of blowdown in small areas. In nearby Menominee County, Wisconsin, 80 million board feet (one board foot = 2,359-$cm^3$) of drought-weakened timber blew down from 1930 to 1937 (Secrest et al., 1941). These data emphasize the importance of wind in the western extension of the northern hardwood forest, but, because of marked differences in topography and wind regimes, it is difficult to relate these experiences directly to the forests of northern New England.

Smith (1946) proposed that New England forests are subject to three major classes of destructive windstorms: tropical cyclones (hurricanes), extratropical cyclones, and more-localized intense winds associated with stormfronts and thunderstorms. Northern New Hampshire and Vermont have been struck by severe hurricane-force winds only twice, in 1815 and 1938, over the period 1492 to 1976 (Ludlum, 1963). Smith would include the 1788 hurricane, but Ludlum (1963) indicates that this storm was confined to southeastern Vermont and central New Hampshire.

The effects of the hurricane in 1938 are reasonably well known because of survey records by the U.S. Forest Service (see Smith, 1946; Spurr, 1956a). Nearly three billion board feet of timber were blown down and 243,000 ha of forest land in Connecticut, Rhode Island, Massachusetts, and Vermont were severely damaged. The total forest land in those states was about 5.3 million ha in 1940 (Barraclough, 1949), of which we estimate about 4.5 million ha were in the zone of hurricane damage mapped by Smith (1946). Thus, only about 5% of the zone affected by the hurricane was severely damaged. Although many stands were completely destroyed or lost patches of trees, most stands within the mapped region escaped with damage to isolated trees. Most observers agree that topography was a major factor in determining damage (Smith, 1946).

At the Bartlett Experimental Forest, approximately 65-km northeast of Hubbard Brook, Victor Jensen, silviculturalist with the U.S. Forest Service, observed that damage to northern hardwood forests from the 1938 hurricane was very limited. It was principally confined to defective trees or local wind-prone sites, while most old-growth and even-aged northern hardwoods were relatively unaffected (Jensen, 1939, personal communication).

The long return period, two severe hurricanes during 484 years, and the restriction of the most devastating effects to particular sites relative to the position of the storm track, suggest that hurricanes are not a major recycling factor for most parts of the hardwood forest of northern Vermont and New Hampshire. Smith (1946) also considered these northern hardwood regions to be areas at low risk from large-scale storms, including extratropical storms.

The intense relatively localized winds associated with frontal showers and thunderstorms may be a more important factor than larger-scale storms, because of their much greater frequency and, perhaps, their greater cumulative areal coverage. There are almost no precise data on these winds, but cursory observations of northern hardwood forests in the Hubbard Brook region over the last 20 years suggest that small-scale windthrow of single trees to small- and moderately-sized patches of trees is not uncommon.

In the presettlement forests of northern Vermont, 5 of 163 comments on vegetation along land survey lines mentioned windfalls of various sizes (Siccama, 1971). Lorimer (1977) also records that windfalls covered 2.2% of the total surveyed distance in northeastern Maine in the 1793–1827 surveys. Most of these windfalls were recorded in conifer forests, while only a few occurred in mixed stands or in hardwood stands.

*Windthrow Mounds and Forest History.* Stephens (1955, 1956), by a careful analysis of windthrow mounds, fallen trees in all stages of decay, and living trees, was able to reconstruct a 500-yr history of a 0.4-ha area in the Harvard Forest in central Massachusetts. About 14% of the area of the plot was covered in mounds and pits (Figure 4-8). The abundance of mounds may have been a factor in the selection of the study site, but this point is not made clear. The analysis indicated that four of the six windthrow events were considered to be major: one in the fifteenth century and three others in 1635, 1815, and 1938. These last three dates correspond to the occurrence of major hurricanes in east central Massachusetts (Smith, 1946; Ludlum, 1963). Henry and Swan (1974) applied Stephens' methodology to a 0.04-ha plot in the Pisgah Forest in southwestern New Hampshire. To optimize the opportunity to reconstruct forest history, the plot was placed in an area where rotting material was abundant. They determined that a severe fire about 1665 resulted in more or less even-aged conifer forest which grew without major disturbance for 262 years. From 1897 to 1938 four wind events, culminating in the 1938 hurricane, completely destroyed the larger trees and initiated new growth. These studies emphasize the frequency and magnitude of wind as a disturbing force in forest ecosystems of central New England, but it should be borne in mind that the amount and frequency of disturbance per unit area of land reported in these studies is very much a function of the study site selected by the investigator.

Nevertheless, in a qualitative sense, windthrow mounds are a widespread phenomenon. Their occurrence in northern hardwood forest in Pennsylvania led Goodlett (1954) to conclude that windthrow of individual trees or small groups of trees was an important factor in maintaining intolerant species in the presettlement northern hardwood forest. Stephens (1956), in an 8000-km journey to the Cumberland and Smokey Mountains, south through the southern Piedmont, west to the Quachita and Boston Mountains, and back to Massachusetts, found mounds and pits of uprooted trees practically everywhere. Lutz (1940) concluded that windthrow is a universal phenomenon in forest regions and that over long periods of time soil under forest stands may be repeatedly subjected to disturbance when trees are uprooted. This conclusion applies quite well to the Hubbard Brook region and is reflected in the pit and mound topography mapped on Watershed 6 (Bormann et al., 1970). We point out, however, that many of the mounds noted in our study and other studies could result from treefall due to endogenous causes, e.g., trees weakened by competition, disease, or senescence that topple in modest winds or from ice or snow accumulations.

It seems reasonable to conclude that presettlement northern hardwood forests in the White Mountains were subjected to frequent disturbances by wind, but this was mostly of a modest scale. However, localized wind-prone sites might have suffered extensive wind damage at fairly frequent intervals. More intense effects might have been localized on particularly wind-prone sites.

## Natural Catastrophic Rotations

Heinselman (1973) proposed the term "natural fire rotation" as the time necessary to burn over an area equal to an area under consideration. For the Boundary Waters Canoe Area of northern Minnesota this was about 100 years. We believe that, in presettlement times, natural fire or cyclonic wind rotations for much of the White Mountain region were many centuries or even a millenium. Our conclusion is based on the convergence of several lines of evidence.

1. In terms of damage from large cyclonic wind storms, the lower elevations of the White Mountain region are among the lowest wind-risk areas in New England (Smith, 1946), while modern

statistical data suggest that the White and Green Mountain regions are, perhaps, the least susceptible areas to fire within the northern hardwood region.

Many factors, such as recurrence of severe drought years, lightning, condition of the vegetation, propensity to accumulate fuel, and human activities, determine an ecosystem's susceptibility to intense forest fires. Although it is difficult to combine these factors into a meaningful index of fire susceptibility, an extensive record of forest-fire activity on national forest lands from 1945 to 1976 (the period of most reliable record), suggests that forests in the White Mountains and Green Mountains of northern New England are among the least burnable (Figure 7-3).

Fires occur in the White Mountain National Forest at the rate of 4.3 per million acres (4.3/405,000 ha) per year, with about one-third of these ignited by lightning. Although this is the lowest ignition rate for any of the 11 national forests in the northern hardwood region, the more interesting fact is that on the average only 8 acres per million acres; (3 ha/405,000 ha) burn each year. This is the smallest area burned per year among the 11 forests. With the exception of the Green Mountain Forest (10 acres/million acres; 4 ha/405,000 ha) and Chequamegon Forest in Wisconsin (43 acres/million acres; 17 ha/405,000 ha), the White Mountain National Forest annual burn rate is 15 to 40 times lower than that of other forests. It is more than two orders of magnitude lower than that of the national forests in lower Michigan, and about 40 times lower than that of the national forests in Minnesota, one of which contains the Boundary Waters Canoe Area—the site of Heinselman's excellent study (1973) of forest-fire history referred to earlier in this chapter. Perhaps still more interesting is the fact that the highest burn rate in the White Mountain National Forest for any year is 56 acres/million acres (23 ha/405,000 ha). This is one to two orders of magnitude smaller than that of the other forests, with the exception of the Green and Chequamegon Forests.

The principal cause of forest-fire ignition in the northern hardwood region is due to the activities of man (Haines et al., 1975). Data on the recreational use of national forests (Figure 7-4) indicate that the White Mountain National Forest is among the most heavily used, with usage more than two times that of the northern forests surrounding the Great Lakes.

Given the heavy use by man, the lowest ignition and annual burn rates, and the lowest area burned in any one year, we propose several possible conclusions: Fire-suppression crews in New England are more efficient, despite the rugged terrain and access of all areas of the forest to recreational users. New Englanders are more careful in the woods even though the bulk of them reside in cities. Or, and this is our opinion, a combination of meteorologic, biologic, and

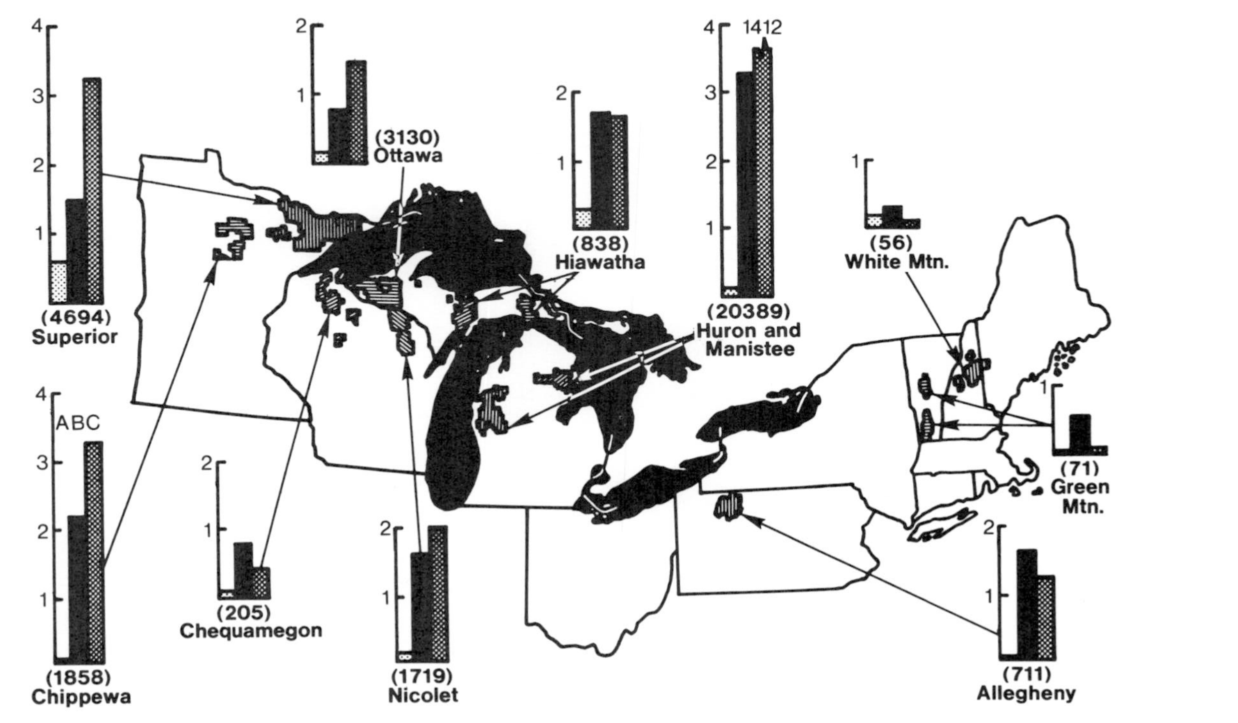


**Figure 7-3.** Fire statistics for national forests in the northern hardwood region. Columns A and B (see Chippewa for key) average annual number of fires, in units of 10, caused by lightning (A) and man (B). Column C average acreage burned annually in units of 100 acres. The number in parentheses is the largest acreage burned in a single year. (All data are per million acres of national forest land.) Data calculated from U.S. Forest Service Annual Fire Reports from the national forests over a 32-yr period, from 1945 to 1976. (Primary data supplied by Junius Baker, Jr., U.S. Forest Service.)

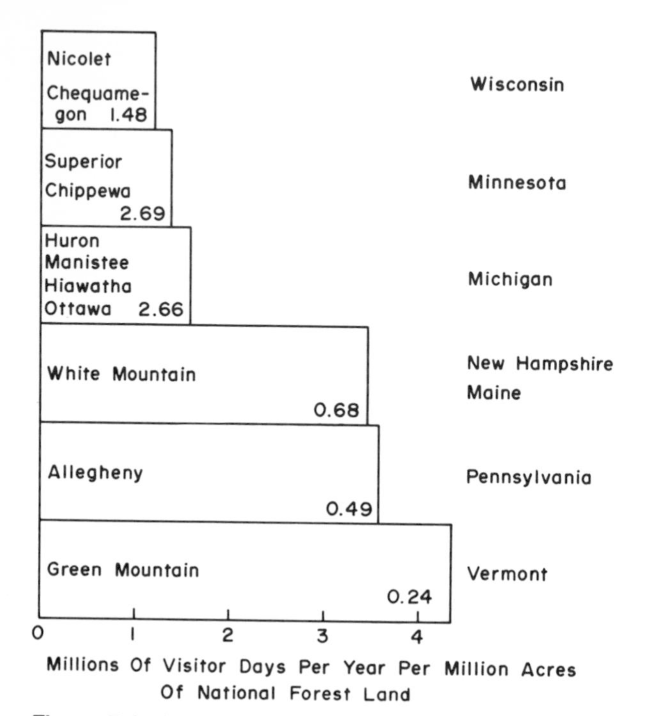


**Figure 7-4.** Average annual recreational use of national forest land in the northern hardwood region from 1965 to 1976 by states. National forests within the states are listed. Weighted average acreage in national forests, 1965–76, in millions of acres is shown by number in bar (1 acre = 0.405 ha). One visitor day equals 12 hr.

topographic factors make White Mountain forests less susceptible to fire.

2. Indian populations, elsewhere a significant factor in the occurrence of presettlement fires, were sparse and had less incentive to burn. At any rate, there is practically no historical evidence for widespread burning during the several centuries preceding the arrival of European settlers. We hope to extend our knowledge of the fire record by charcoal analysis (Swain, 1973) of a 13,000-yr-old sediment core from Mirror Lake in the Hubbard Brook Valley and from peat cores from high-elevation bogs in the White Mountains (M. Davis, personal communication).
3. All areas within the region are not equally susceptible to catastrophic damage. Burning potential or wind susceptibility is determined by a number of physiographic and biologic characteristics. The potential for fire to spread is determined by factors such as patterns of wind flow, distribution of lakes, streams, and wetlands, rugged topography and bare ridges, and droughtiness of soil due to texture, soil depth, slope, and aspect (Heinselman, 1973; Rowe and Scotter, 1973). Wind-susceptible sites are places where high winds

are constricted and accelerated such as along the flanks of hills, over ridge tops, through gaps, or along narrow valleys or soils where deep root penetration is not possible (Smith, 1946). In the well-watered rugged landscape of the White Mountains it seems reasonable to think that many areas would be protected from the easy sweep of fire or from intensive cyclonic winds.

4. Ecosystems themselves contribute to their spatial and temporal vulnerability to wind and fire. Differences in flammability are related to species composition and the accumulation and spatial distribution of organic fuel (Bloomberg, 1950; Mutch, 1970; Heinselman, 1973; Rowe and Scotter, 1973). Similarly, susceptibility to wind damage may be related to size of individual trees, whether they are deciduous or evergreen, and structure of the stand (Smith, 1946; Spurr, 1956a; Henry and Swan, 1974). Although we have not found specific information on the susceptibility to fire of the various developmental stages of northern hardwood forests, a number of ecologists and foresters are of the opinion that mesic northern hardwood forests are relatively resistant to fire (Chittenden, 1905; Bromley, 1935; Hough, 1936; Egler, 1940; Stearns, 1949; Winer, 1955). In the words of many workers in forest fire control, they are "asbestos" forests. On the other hand, it would seem that as northern hardwood stands become composed of larger and older trees their vulnerability to wind increases. Toppling of large trees might be coupled with an increase in susceptibility of the area to fire (Hough, 1936; Stearns, 1949; Wright and Heinselman, 1973).
5. Finally, a basic difference between fire-dependent and fire-resistant ecosystems relates to the rate at which biological decomposition reduces the fuel load of dead wood or organic matter on the mineral soil surface (Wright and Heinselman, 1973). Apparently, in some systems litter production is so much greater than decomposition that fuel accumulation reaches enormous proportions as autogenic succession proceeds (Bloomberg, 1950). In humid temperate or tropical systems characterized by intense fungal, bacterial, and detritivore activity, decomposition can reduce susceptibility to fire by a rapid turnover of litter and a tendency toward modest standing crops of dead wood and forest floor material. Our studies indicate that the forest floor under northern hardwood forests in the White Mountains does not continuously accumulate but rather reaches an equilibrium and that massive inputs of branch and limb material are turned over in one or two decades (Chapter 1). It is interesting to note that, in northern hardwood forests where geologic conditions lead to more base-rich fertile soils (mull soil, Bormann and Buell, 1964), the forest floor (the layer of the organic matter on top of the mineral soil) is completely replaced by bare mineral soil with only seasonal litter cover. Northern hardwood forests with mull soils would be even less susceptible to fire.

## Probability of a Steady State

Loucks' (1970) hypothesis that forest ecosystems are periodically restarted or recycled by fire at intervals short enough so that an autogenically derived steady state is rarely achieved seems reasonable for large areas of fire-dependent vegetation (see *Quarternary Research* **3**(3), 1973). Before discussing the applicability of this hypothesis to the White Mountains region, it seems appropriate to point out that Loucks' concept of catastrophic recycling and our concept of steady state share some basic assumptions. Both include disturbance and "restarting" as fundamental components of ecosystem function, and in some respects the difference between the two concepts is best considered one of timing and degree. Our steady state includes endogenous disturbance and constant recycling by randomized local perturbations which eventually cover the whole area. Loucks' catastrophic perturbations are applied to the whole system at once, and the whole system is recycled in one time sequence; in essence, the patches composing the system are synchronized.

The evidence available to us indicates that Loucks' hypothesis cannot be generally applied to northern hardwood ecosystems of the White Mountains. During presettlement times, there may have been some fire-prone areas in which return periods of fire were short enough to regularly truncate autogenic development, or over millenia the whole region might have been recycled. Using modern fire records for the Province of New Brunswick, Canada, Wein and Moore (1977) have estimated fire-rotation periods (i.e., 100 divided by the percentage of total area burned annually for a forest type) ranging from about 650 to 2000 years for various forest types in which northern hardwood species are important. Similar estimates derived from the modern fire record for the White Mountain National Forest (Figure 7-3) would yield incredibly long fire-rotation periods. Consequently, it seems that fairly large areas of presettlement hardwood forest within these humid, rugged, and highly dissected mountains were free from catastrophic disturbance for periods of time sufficient for a steady-state condition to have been achieved.

The presettlement northern hardwood landscape might be conceived as a collection of ecosystems in various stages of development. Some areas, particularly fire- or wind-prone sites, might have been regularly recycled by catastrophic disturbance, but substantial areas might have supported ecosystems approaching the steady state. We say "approaching" because smaller-scale exogenous disturbances by wind, snow, or ice that kill or maim healthy trees seem unavoidable over extensive areas for long periods of time. These disturbances might best be considered within the pattern of autogenic development, but their imposition on the ecosystem would upset the theoretical proportions of States A, B, and C based on endogenous disturbance alone (Figure 6-5).

The "defense" of the reality of the autogenically derived steady state may seem to be a polemic exercise, but it serves two useful purposes. Ecology, like all other fields, is affected by the swinging pendulum of opinion. For decades Clementsian notions of succession and climax thoroughly dominated much ecological thinking and buttressed land-use policies that assigned disturbances a minor role in most naturally occurring ecosystems (Raup, 1967). Quite properly, these ideas have come under attack, and the concept of catastrophic disturbance, particularly fire, as an integral part of the structure and function of temperate forest ecosystems has gained ground. But the pendulum may swing too far toward catastrophic recycling at the expense of concepts of long-term autogenic development that are themselves continuing to evolve. This is not a trivial matter, for theories of ecosystem development are beginning to play a major role in studies of evolution (e.g., Loucks, 1970; Connell and Slatyer, 1975; Smith, 1975) as well as in land-use management decisions (e.g., Kilgore, 1973; Wood and Botkin, unpublished data).

Finally, we suggest that it might be useful to consider the catastrophic recycling hypothesis and the autogenically derived steady state as components of a two-dimensional geographical space. Because of modest levels of exogenous disturbance or long periods between large-scale disturbances, forested regions of the humid, temperate mountains of the northeastern United States might be considered as centers where autogenic development has its highest probability of achieving steady-state conditions, while in almost every direction away from these centers catastrophic recycling takes on increasing importance.

## POSTSETTLEMENT DISTURBANCE

Settlement by Europeans markedly changed the nature, frequency, and extent of exogenous disturbance in the northern hardwood forests of the White Mountain region. Changes were primarily affected by land-use practices, lumbering and land clearing, and a marked increase in the frequency and intensity of fires associated with these practices, and with a general increase in population size within the region. Brown (1958a,b) proposes three periods of land utilization for the area around Hubbard Brook. These apply generally to the White Mountain region but with somewhat different dates.

### The Period of Settlement: 1760–1840

Land clearing for agriculture and production of wood products were major activities during this period. However, most attention was devoted to subsistence farming, and the production of lumber was limited to local needs. Except for river valleys and lower elevation tracts, Brown believed that these early settlers had relatively little effect on the primeval forest. This idea is supported by data on land clearing in the town of Thornton, adjacent to the Hubbard Brook watershed (Table 7-1), where only 6% of the total land area was cleared by 1830.

**Table 7-1.** Land Usage During 1790–1830 in Thornton, New Hampshire[a]

| Usage | 1790 | 1800 | 1816 | 1820 | 1830 |
|---|---|---|---|---|---|
| Mowing | 96 | 148 | 192 | 190 | 202 |
| Pasture | 68 | 152 | 260 | 332 | 488 |
| Arable | 30 | 37 | 53 | 51 | 81 |
| Orchard | 0 | 0 | 9 | 13 | 18 |
| Percentage of total area | 1.5 | 2.6 | 3.9 | 4.4 | 6.0 |

[a] *Values are in hectares. Thornton has a total area of 13,209 ha (from Likens, 1972b).*

## Farm and Forest Economy: 1840–1880

During this period, land clearing reached its maximum and began to decline, but the tides of clearing, cultivation, and abandonment flowed unevenly. New lands were being cleared, while others were being abandoned. There are no accurate statistics on the total area of northern hardwood land cleared in the White Mountains (Likens, 1972b). One of the most important events was the penetration of the railroad into these mountainous reaches. Transportation of bulk goods at low prices made possible the utilization of previously little-recognized timber resources. In addition to providing an opportunity to transport lumber and logs to distant markets, the railroad directly increased the use of wood. For approximately 30 years, 1850–1880, many locomotives were wood burners. Lumbering and sawing and the agricultural life of the community became closely dovetailed. Farming and saw milling were spring and summer occupations, while attention was focused on cutting and getting logs out in the winter. Logging was generally limited to small, irregular, and scattered units. The highly diversified ownership pattern and Yankee individuality discouraged aggregation of the large units necessary for larger timber operations; however, a few large professional lumbering companies emerged during this period (Coolidge and Mansfield, 1860). Most cutting occurred fairly close to the railroad or to rivers large enough to support log drives, and the relative abundance of timber elsewhere in New England held stumpage values at such a low level that the policy of selecting only the larger and better trees was an economic necessity. Spruce was the favored species and was often selectively cut some distance into the mountains. Hardwoods were rarely used as lumber. At the close of this period, large areas of the White Mountains were still "virgin forest" (Brown, 1958a; Gove, 1968).

## Large-Scale Commercial Logging: 1880–1940

Large-scale forestry operations in northern New Hampshire date mostly after 1870. Until 1869, the state of New Hampshire owned the greater part of the White Mountain region and Coos County north of it. However, the policy of the state was to dispose of public lands as fast as possible, and large tracts were sold for almost nothing (Chittenden, 1905). By 1890, because of increased demand, better prices, increased capitalization, and better transportation, lumbering had become one of the great industries of the state (Walker et al., 1891). One of the major factors leading to large-scale commercial logging was that numerous logging companies were able to acquire units of land sufficiently large to sustain logging operations for a period of years. With the land base secured, it was possible to invest the large quantities of capital needed to build the necessary internal transportation facilities, i.e., roads and logging railroads. To defray the substantial overhead costs, it was thought necessary to clear-cut. At first trees were cut for lumber, but with time pulp became increasingly important. Softwoods were preferred, but, after the development of hardwood pulping methods in the early 1900s, hardwoods were utilized as well (Brown, 1958b).

The *Report of the Forestry Commissioners of 1893* estimated that all timber of any commercial value had been removed from 0.45 million ha in a 120 × 175-km rectangle that included the White Mountains, but that about 160,000 ha of primeval forest remained (Walker et al., 1891). Timber of commercial value refers to softwoods, since hardwoods had practically no sale value at that time (Chittenden, 1905).

The Forest Survey of 1903, fairly detailed for its day, reported that about 12% of northern New Hampshire supported virgin spruce forest but that there were practically no virgin hardwood stands. Most hardwood stands had contained some spruce, but the spruce had been removed by high-grading (Chittenden, 1905). By 1940, most of the area had been cutover and few virgin stands remained (Brown, 1958b).

## Postsettlement Forest Fire

The incidence and extent of forest fires in the White Mountain region greatly increased during postsettlement times. For the early settlement period (the late eighteenth century), there is evidence that extensive fires occurred in southern New Hampshire and in Maine (Plummer, 1912; Fobes, 1948; Lorimer, 1977). These fires were often associated with the burning of slash during land clearing and ignition of slash after lumbering.

Forest fires were apparently common throughout New Hampshire in the post Civil War period. The Forestry Commission commented in 1885 that, in the previous 25 years, forest fires destroyed as much timber as had been harvested (Hale et al., 1885). They also commented that the primeval forest within the White Mountains had recently been invaded by the axe and that there was every reason to suppose that fire would quickly follow. In 1891, the Forestry Commission reported (Walker et al., 1891) that extensive tracts within the White Mountains had been burned. These fires included one of the largest fires recorded for the region, the Zealand fire of 1888, which covered 5000 ha (Chittenden, 1905). This fire occurred in 1886, according to Belcher (1961). In 1893, the Forestry Commission again emphasized that fires follow cutting and that standing forests do not burn well.

Chittenden (1905), in what was probably the most reliable survey of the time, mapped burned areas throughout the northern third of New Hampshire. About 6% (34,000 ha) of the total area was burned in scattered fires in 1903, an exceptional fire-year in the northeastern United States. Chittenden reported that the virgin forest was strikingly free of fire, despite the exceptional fire-year. He also offered the opinion that the mountainous topography precludes fires that sweep unhindered over vast areas.

Since the advent of fire protection, fire is of relatively rare occurrence in the White Mountains region, despite the fact that the region is heavily used by tourists, campers, and hikers, as well as by wood-using industries. As stated earlier, for the 30-yr period prior to 1976, only 0.025% of the 275,000 ha comprising the White Mountain National Forest had been burned.

### Effects of European Man

During presettlement and postsettlement periods, European man increased the frequency and areal extent of major exogenous disturbances in the northern hardwood forests of the White Mountains. Virtually the entire area has been cut at least once, either by clear-cutting or high-grading, many areas have been burned, and substantial areas have been cleared and abandoned.

It is impossible to make a general statement about the effects of increased catastrophic disturbances on the structure and function of the northern hardwood ecosystem. Some potential for future growth was probably lost in severely burned or eroded areas (Chittenden, 1905), and a few systems were driven over the stability–resilience border (Figure 6-9). Today the hardwood area is mostly covered by young forests, many of which are now undergoing cutting for a second or third time.

## INCREASED REGULARITY OF WHOLE-SYSTEM BIOMASS OSCILLATION

The intervention of European man in northern hardwood forests appreciably increased the land area occupied by even-aged forests. In presettlement forests, relatively large areas were probably in an all-aged condition owing to autogenic processes working over long periods of time without interruption by large-scale catastrophic events. Many stands were probably composed of patches in all stages of development, i.e., they were asynchronous.

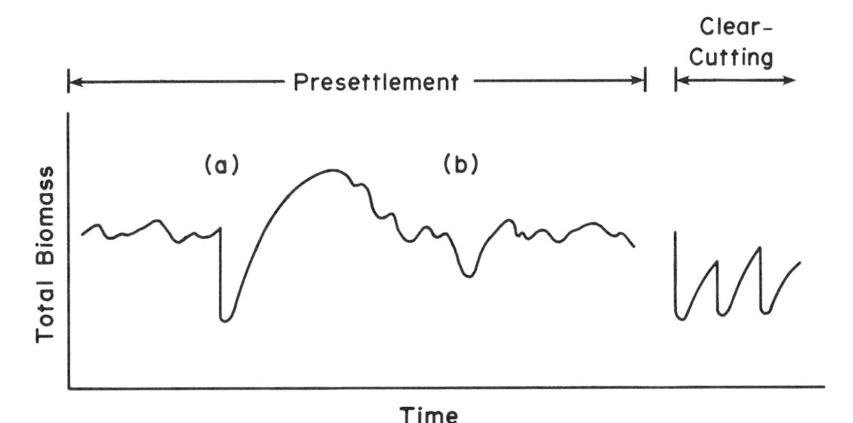


**Figure 7-5.** Hypothetical pattern of total biomass in a northern hardwood ecosystem in the presettlement period, when system-wide perturbations [(a), fire; (b), wind] recur at fairly long intervals, and in the modern period, with regular harvest by clear-cutting at 60- to 70-yr intervals.

Activities (such as clear-cutting, intense fires, or land clearing and abandonment) that remove preexisting living biomass often result in the establishment of even-aged stands (Gilbert and Jensen, 1958). In effect, these activities might be thought of as organizing forces that impose synchrony on the patches composing the ecosystem. The development of an initially even-aged synchronous stand after clear-cutting represents the central theme of this book.

The second- and third-generation cuts now occurring in the northern hardwood forest tend to be some form of clear-cutting (blocks, strips). This, of course, truncates the trend of the developing forest as it proceeds toward asynchrony and reestablishes synchrony (Chapter 6). Thus, continuing harvest of the hardwood forest by clear-cutting has the effect of increasing the regularity with which the biomass of the whole ecosystem oscillates (Figure 7-5). Management of the northern hardwood forest by clear-cutting produces a pattern of recycling through time quite similar to that proposed for the natural behavior of ecosystems in fire-prone areas (Loucks, 1970). Paradoxically, in the latter areas, management (fire suppression) is disrupting the short-term recycling thought to be characteristic of those areas in presettlement times and is leading toward an autogenically derived condition seldom achieved previously (Heinselman, 1973).

## Summary

1. Disturbance is defined as the disruption of the pattern of the ecosystem by physical forces; fire, wind, and the force of falling trees.
2. Endogenous disturbance is an integral part of the autogenic developmental process.

3. Major exogenous forces, fire, wind damage, and clear-cutting, all result in an immediate decrease in gross primary productivity but are dissimilar in their effects on the oxidation/GPP ratio, nutrient and biomass export, and on the age structure of the ecosystem immediately after the disturbance.
4. Fire during presettlement times was common in most areas surrounding the White Mountains of northern New Hampshire but relatively uncommon within the mountains.
5. Large-scale cyclonic winds are not a major disturbance factor over much of the lower elevations in the White Mountains, but small and intermediate wind effects may be important agents of disturbance over the long term.
6. Large-scale disturbance by natural catastrophes in the presettlement northern hardwood forests of the White Mountain region is thought to occur only at long intervals and is not considered to be an inexorable deterrent to the development of a steady-state condition.
7. Care should be exercised in extrapolating hypotheses of catastrophic ecosystem recycling developed in one region to other regions. The forested region in the humid mountains of northern New England may be considered a center where autogenic development has its highest probability of achieving steady-state conditions, while in almost every direction away from this center catastrophic recycling takes on increasing importance.
8. During postsettlement times, European man has greatly increased the frequency and extent of exogenous disturbance in the White Mountain region. This has been done through forest cutting, land clearing and abandonment, and through an increased incidence of fire.
9. Management of northern hardwood forests by clear-cutting produces a pattern of recycling similar to that proposed for fire-prone areas during presettlement times. Conversely, fire suppression in the latter areas leads toward an autogenically derived condition seldom achieved in presettlement times.

# CHAPTER 8

# Forest Harvest and Landscape Management

When designing a plan to harvest forest products, forest managers face a variety of questions: Will the plan yield the most satisfactory profit commensurate with short- and long-term expectations? Does the plan promote or allow for the establishment of an adequate crop of new individuals of desirable species? Is the environmental impact acceptable? How can a high level of productivity of the forested ecosystem be sustained? What is the maximum or optimum rate of harvest?

These same concerns are voiced by conservationists with somewhat different emphasis: What are the short- and long-term impacts of various harvesting practices on natural forests, including soil fertility and the quality of drainage waters? How rapidly will the landscape return to its natural condition? Will the natural system be degraded by repeated harvest? Is the area being used or managed in the best way for all interests?

The Hubbard Brook Ecosystem Study can contribute partial answers to some of these questions as they apply to northern hardwood forests, but perhaps the most useful contribution is to allow a more precise formulation of the questions themselves.

## AIR POLLUTION

Traditionally, landowners have been concerned about the protection of their property against destruction from outside forces, e.g., vandalism, fire, and, more recently, off-the-road vehicles and camping. However, there are more subtle outside influences, such as air pollution, which may seriously degrade landscapes. These factors should be of major concern to landowners, managers, and conservationists. Some of the important questions include: What areas are affected and to what degree? What are the spatial and temporal dimensions of the problem? What are the short- and long-term ecological effects of air pollutants on forested and associated aquatic ecosystems? What and where are the sources of the air pollutants, and how can they be controlled? How do the costs of pollution control compare with the benefits to and rewards from the affected natural ecosystems? We believe that owners of forest land and conservationists should enter the current debate about standards for air pollution and should work toward greater protection of forest land from regionally dispersed air pollutants.

What are some of the problems, real and potential? Our work at Hubbard Brook has shown that a variety of materials, injected into the atmosphere from man's activities, are dispersed over large regions and transported by air masses to areas quite remote from large urban or industrial centers. Airborne gases and particles may be removed from the airstream in several ways: by gravity in wet and dry deposition, through impaction on surfaces, and by biological assimilation of gases. All three of these mechanisms operate in humid forest ecosystems, so that relatively large quantities of air pollutants may be removed from the air in these areas. Strong acids (e.g., $H_2SO_4$), oxidants (e.g., ozone), heavy metals (e.g., Pb), and pesticides (e.g., DDT) are some of the anthropogenic airborne contaminants that are spread over wide areas (Likens and Bormann, 1974a,b,c; Smith, 1975; Bormann, 1976; Braekke, 1976; Cleveland et al., 1976; Dochinger and Seliga, 1976; Groet, 1976; Likens, 1976).

## Acid Precipitation

Acid precipitation (rain and snow at a pH $<5.6$) is now common over most of the eastern United States (Chapter 2; Likens and Bormann, 1974; Likens, 1976) and is affecting forested landscapes. Forested ecosystems are very complex and are affected by a variety of environmental variables. Thus, it is difficult to ascertain the specific effect of acid precipitation within this milieu. The added stress, nevertheless, is real. Field and laboratory studies have shown that acid precipitation can increase the leaching of nutrients from the foliage (Wood and Bormann, 1974; Abrahamsen et al., 1976), accelerate cuticular erosion of leaves, produce leaf damage when the pH of precipitation falls below 3.5, alter response to associated pathogens, symbionts, and saprophytes, affect the germination of conifer seeds and the establishment of seedlings, affect the availability of nitrogen in the soil, increase leaching of ions from the soil, decrease soil respiration, and lower rates of decomposition (*Ambio*, 1976; Braekke, 1976; Dochinger and Seliga, 1976; Likens, 1976). In fact, forest growth in the eastern United States appears to have declined since about

1950 (Whittaker et al., 1974; Cogbill, 1975), but it is impossible to state at this time what the role of acid precipitation has been in this decrease (Cogbill, 1976).

## Heavy Metals

Using the moss *Leucobryum glaucum* as a collecting agent, Groet (1976) determined that about 24-mg of lead/$m^2$-yr were deposited in the vicinity of the White Mountains. Schlesinger et al. (1974), on the basis of bulk precipitation samples, calculated a value of about 20-mg of Pb/$m^2$-yr for the same general area. These and other data suggest that relatively high rates of deposition of airborne heavy metals are occurring in the northeastern United States. Increased amounts of Ag, Au, Cr, Ni, Pb, Sb, and V in recent lake sediments in the Adirondack Mountains of New York State and the White Mountains of New Hampshire also have been attributed to long-range atmospheric transport and deposition (Galloway and Likens, 1977). The biological consequences of the pollutants that are currently falling on the northeastern United States are a major concern.

The forested landscape is a very efficient filter. Many airborne and waterborne substances are effectively removed from active circulation by deposition in woody biomass, forest soils, or sediments in lakes or other impoundments. One striking example is lead. Some 97% of the lead falling on the forest at Hubbard Brook is retained and stored within the ecosystem (Siccama and Smith, 1978). Drainage waters from the Hubbard Brook Experimental Forest are of high quality, whereas concentrations of lead in precipitation frequently exceed published limits for drinking water. We must ask: How long can ecosystems carry on these filtration activities without serious impairment of their natural structure, metabolism, and biogeochemical functions?

## Oxidants

Ozone from photochemical air pollution, emanating from the New York City area, has been traced to sites in Connecticut and northeastern Massachusetts. In fact, southwestern Connecticut is reported to have one of the worst ozone pollution problems in the country, with daily maximum concentrations frequently more than double the Federal standard of 0.08-ppm (Cleveland et al., 1976). The effects of oxidants such as ozone on the northern hardwood ecosystem are poorly known. Western forests have been seriously damaged by ozone and peroxyacetyl nitrate (PAN) in various areas (e.g., Miller, 1973; Miller and Elderman, 1977; Williams et al., 1977).

The added stress of photochemical oxidants on the northern hardwood ecosystem, in combination with acid precipitation, heavy metals, and other airborne pollutants, may represent a threat to the continued productivity of these forests.

# FOREST HARVESTING PRACTICES

## Clear-Cutting Methods

It is not our objective to review and evaluate all types of timber-harvesting practices. Rather we comment on only three types of clear-cutting: *stem-only harvesting*, in which all merchantable trunks between stump and crown are harvested; *whole-tree harvesting*, in which the entire portion of the tree aboveground is harvested and chipped for use in paper pulp or reconstituted paper products; and *complete-tree harvesting*, in which the entire tree, aboveground and belowground, is harvested and chipped. Currently, in the White Mountain region, stem-only harvesting, with lumber, millwood, and pulpwood as end products, is the most widely used. Whole-tree harvesting on a commercial scale has just begun (Bryan, 1976). A special variety of complete-tree harvesting which involves 10- to 20-yr harvests of puckerbrush (the mixture of woody species that arises after a clear-cut) is receiving considerable study at the University of Maine (Chase et al., 1971, 1973; Ribe, 1974; Young, 1974).

## Regeneration of Vegetation After Clear-Cutting

Our studies of recently clear-cut sites, cut by the stem-only method, confirm what foresters have known for some time. The regenerative capacity of the northern hardwood forest ecosystem in the White Mountain region is excellent as long as cutting is done in a careful manner (Chapter 4). Regeneration after cutting results from activation of a variety of reproductive and growth strategies and from a temporary increase in resource availability (available nutrients, solar radiation, and water) after cutting. Together these lead to rapid regrowth of vegetation (Chapter 5). Erosion, which is generally not greatly accelerated with careful cutting methods, is rapidly brought under control. The major biological effect of clear-cutting is to promote and synchronize regeneration and growth mechanisms (Table 4-1) that usually occur naturally in scattered openings in the forest.

As yet there is little information on regeneration after harvesting by the whole-tree or puckerbrush methods. However, whole-tree harvesting results in considerably more soil disturbance than stem-only harvesting (Zasada, 1974), but this point is debatable (R. Pierce, personal communication). Complete-tree harvests involve tearing stumps out of the soil. Increased disturbance of the forest floor and/or removal of living root systems might inhibit regeneration of those species dependent on advance regeneration, buried seeds, and vegetative reproduction.

## Nutrient Export After Clear-Cutting

In general, in the White Mountain region, there is a significant increase of dissolved nutrients in drainage water after the stem-only type of clear-cutting. Although this may be thought of as a cost paid for rapid recovery (Chapter 5), it should be given careful consideration in the design of the harvesting method. Loss of dissolved substances not only represents loss of nutrients from the ecosystem but also has an environmental impact on aquatic ecosystems draining the cutover forest. As long as clear-cuts are limited to small areas, e.g., several hectares, and the cut area forms a small part of a larger forested watershed, biotic activities in streamwater (Fisher and Likens, 1973; Likens et al. 1978; Meyer, 1977) plus dilution by water from uncut forests (Hornbeck and Federer, 1975) will tend to minimize environmental impacts. However, careless cutting, cutting procedures that reduce the effectiveness of regenerative mechanisms or accelerate losses of dissolved substances and particulate matter, or cutting of too large a proportion of the total drainage basin could promote significant downstream environmental impacts.

In designing harvesting procedures, thought should be given not only to the immediate increase in loss of nutrients in drainage waters resulting from new cutting but also to the rate and amount of drainage loss from previous cuts in the same watershed. In evaluating the total environmental effects, the additive impact of all cuts—past and present, large and small—must be considered.

## Management of the Forest Floor

The forest floor plays a central role in the biogeochemistry and ecology of northern hardwood ecosystems on acidic geologic substrates (Chapter 2). Mineral soils in the White Mountains tend to be relatively infertile, with a relatively high proportion of available nutrients localized in the forest floor. Not only does the forest floor play a major role in annual nutrient cycling, but it acts as a regulator of ecosystem metabolism by storing nutrients during the Aggradation Phase and releasing them after disturbance (Chapters 3–5). The forest floor carries out other functions as well: It promotes infiltration and percolation of water, minimizes erosion, and serves as a repository of buried seeds and the rooting substrate for advance regeneration.

*Convergence.* The forest floor is a product of ecosystem development and of convergence (Dansereau, 1957). Convergence results from the tendency of nature to moderate extremes and to move toward a middle position. For example, the long-term development of a northern hardwood ecosystem (i.e., over a millenium) may result in an absolute increase in resource availability through formation of the soil profile, but development also is marked by the tendency for the characteristics of heterogenous soil microsites to even out or converge. Thus, wetter and drier microsites become more mesic as development proceeds, and it seems likely that changes in soil moisture would be accompanied by convergent changes in soil fertility. Of course, convergence cannot overcome all initial differences in microsites (also, new microsites are continuously being formed, e.g., by treefall), but convergence would tend to moderate extremes within the ecosystem and to even out the spatial distribution, or geometry, of resource availability. Countless cycles of leaf litter might be considered a major, but imperfect, mechanism whereby organic matter and nutrients are distributed fairly evenly in the forest floor. This, in turn, leads toward a more equable distribution of primary production. In other words, productivity of the ecosystem may be affected by the geometry of the forest floor.

*Seedbed Preparation.* Manipulations of the cutover site to expose mineral soil and encourage the establishment of light-seeded species might work in a direction opposite to that of nature by rearranging the geometry of forest floor material.

Rearrangement of the forest floor may result from a variety of activities such as road building (discussed below), log skidding, and seedbed preparation. The creation of skid trails is sometimes encouraged to enhance the establishment of birch seedlings (Filip, 1969; Kelso, 1969). In some instances, special instruments such as bulldozers equipped with blades, rock rakes (Filip and Shirley, 1968), or disks (Bjorkbom, 1967) have been used to expose mineral soil.

In preparing seedbeds, these questions should be considered:

1. Does stirring and mixing of the forest floor promote more rapid decomposition and mineralization while at the same time diminishing the natural regenerative response of the ecosystem? Experimental results with forest floor material from Hubbard Brook show that mixing promotes an increase in both ammonification and nitrification (Melillo, 1977), while Chase et al. (1968) report that mixing organic and mineral layers stimulates both decomposition and nitrification. If this occurs on a practical scale, seedbed preparation might lead to greatly increased losses of dissolved substances in drainage water.
2. Does the method of seedbed preparation severely disturb the forest floor, diminish control of erosion and transportation, and significantly accelerate loss of particulate matter?
3. Is there a cost in diminished productivity of the resulting stand associated with seedbed preparation techniques, which increase disruption of the forest floor to foster establishment of light-seeded, commercially important species? In studies of seedbed preparation, numbers of seedlings established on a prepared site relative to seedlings established on the undisturbed forest floor often are used as the major criterion determining the success of the treatment. This

approach may underestimate the possibility that the factors which promote the establishment of seedlings may not be optimal for subsequent growth. Yet, production is the ultimate economic goal of management. This problem has several dimensions.

a. There is some question as to whether seedbeds in mineral soil favor the very species (i.e., birches) they were created to promote. Numerous workers (Tubbs, 1963; Marquis et al., 1964; Hoyle, 1965; Marquis, 1965, 1973) have reported superior growth of birch in humus rather than mineral soil, even though germination and survival were greater on mineral soil (Marquis et al., 1964). These workers speculated that differences might be due to greater nutrient availability in humus.
b. Earlier we raised the possibility that seedbed preparations may increase dissolved nutrient losses from the ecosystem. Such losses might reduce subsequent productivity of the exosystem (Lundmark, 1977).
c. Site manipulations that lead to extremely irregular distributions of forest floor material, with substantial areas of exposed mineral soil and areas of heaped-up organic matter, might result in a geometric distribution of resource availability that is less favorable to overall productivity than the natural distribution.

## Roads

The construction of both permanent roads and temporary logging roads has received considerable attention from forest engineers and foresters (e.g., Kochenderfen 1970; Stone, 1973; Patric, 1976a,b; Swanston and Swanson, 1976). It is not our intention to review the extensive literature associated with road building in forested land and its ecological impact. Rather, we discuss briefly an area of research that our study of the northern hardwood ecosystem suggests might be of widespread interest, or at least one that should receive additional thought from forest managers and ecologists alike.

A major emphasis among those studying the environmental impact of road building has been the effects upon hydrology and various soil-erosion processes (Swanston and Swanson, 1976). With the exception of a handful of papers (e.g., Pfister, 1969), the long-term effects of road building on future forest productivity have received little attention. Yet judging by the report of the Chief of the U.S. Forest Service (*Report of the Chief of the Forest Service: 1970–75*) on the number of roads constructed annually on U.S. Forest Service lands alone, it seems safe to assume that tens of thousands of kilometers of permanent and temporary roads are constructed annually on public and private forested lands in the United States. In the northern hardwood forests of the White Mountain region, about 2 to 10% of the harvested area (J. Hornbeck, personal communication) is utilized in the construction of temporary logging roads.

Patric (1977) cites 6 to 16% in West Virginia, while Kochenderfer (1977) reports that roads and landings occupy 10% of nine central Appalachian areas logged with wheeled skidders and 8% in areas logged with jammers.

Temporary roads may alter long-term productivity in a number of ways (Figure 8-1), all of which should receive further study. One of the most obvious impacts of road building in the northern hardwood ecosystem is disruption of the forest floor geometry already discussed. Forest floor material is removed from the area occupied by the road crown, side ditches, and any adjacent slopes used in cut and fill (Akesson, 1963); downslope forest floor material also may be covered by fill. Thus, forest floor material may be removed from or buried over widths ranging from about 9 m in fairly level areas (Akesson, 1963) to greater widths as slope increases. We should assess whether the rearrangement of forest floor material in road building results in changes in productivity; for example, will accelerated growth of trees in areas of humus deposition offset lesser growth in areas of exposed mineral soil?

A much more subtle question concerns the rearrangement of subsurface flow patterns. In addition to the obvious flow of water in stream channels, both permanent and temporary, water moves downslope as subsurface flow. Subsurface flow helps to maintain flow in stream channels during both wet and dry periods. In the summer of 1976, we dug

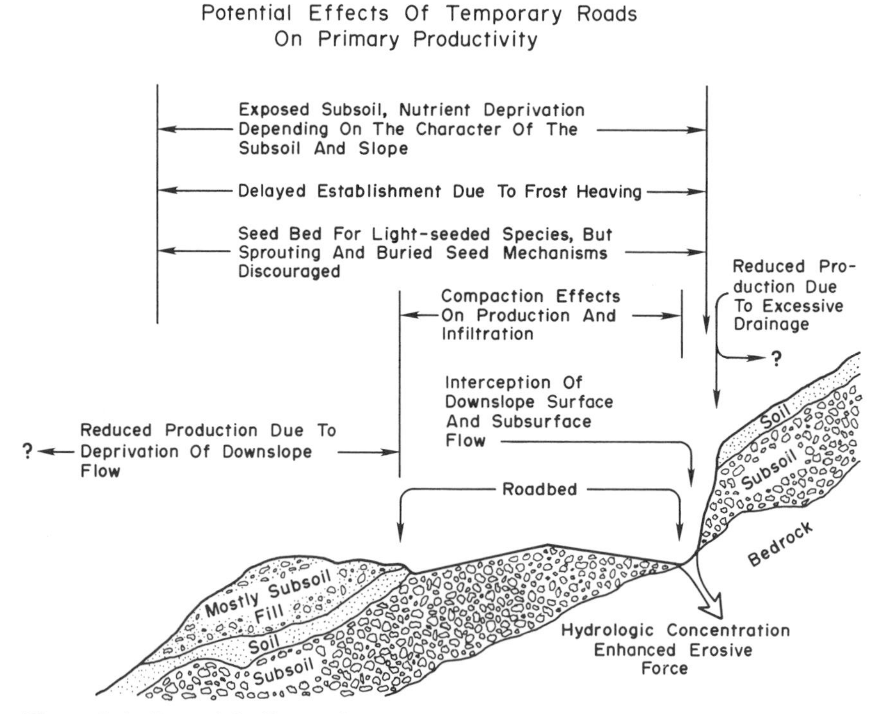


**Figure 8-1.** Potential effects of temporary roads on primary productivity.

a 22-m-long soil pit along a contour in the Hubbard Brook Experimental Forest. The pit penetrated into the underlying impervious till. Inspection of this pit after a heavy rain indicated seepage above the till throughout the downslope face of the pit. In one place, although there was no surface evidence of subsurface drainage, i.e., vegetation characteristic of wet sites, there was an underground channel carrying flowing water.

We should evaluate the role of the downslope movement of surface water in the maintenance of ecosystem productivity: Does subsurface flow provide the vegetation of lower slopes with a supply of water in addition to precipitation, and does it contribute to the generally high levels of production typical of lower slopes?

Megahan (1972) has studied the effects of logging roads on subsurface flow in the mountains of central Idaho. He estimated that roads in his study area intercepted 10 to 20-cm of downslope flow, concentrated it in drainage ditches, and channeled it to streambeds. Megahan (1972) and E. R. Burroughs (personal communication) have pointed out that the conversion of subsurface flow into surface flow by roads may have profound ecological effects on downslope vegetation. We suggest that this could be a fruitful area of research in northern hardwood forests of the northeastern United States, as well as in the northwestern United States.

## Long-Term Effects of Forest Harvesting

Perhaps the most difficult question for the forest manager is how to evaluate the long-term effect of harvesting procedures on the future structure, metabolism, and biogeochemistry of the forested ecosystem. This question has tended to polarize the thinking of a number of foresters (Stone, 1973; Marquis, 1976; Patric, 1976a,b) and conservationists (Curry, 1971), who cite either the continuous use of some European forests or the occurrence of forest catastrophes to prove that harvesting may or may not have a long-term impact on the forested ecosystem.

Actually, the question of maintenance of long-term forest productivity, except for instances of fairly obvious degradation associated with harvesting, is very difficult, if not impossible, to answer with any precision using historical data. The complexity of the forested ecosystem, the fickleness of the climatic variables affecting growth, the time periods involved, and the incompleteness of historical records on growth variables like primary production preclude the detection of any but the most gross changes in productivity due to repeated cycles of harvesting.

Our inability to measure the long-term impact of repeated harvesting on most forest lands points up a major need—the construction of realistic forest growth models that would provide reasonable estimates of the long-term ecologic and economic cost/benefit relationships resulting from various harvesting procedures. Until such understanding is developed, estimates of long-term effects will have to remain largely what they are—educated guesses. Given the complexity of forest ecosystems, basing policy on educated *impartial* judgments is not as unscientific as it may seem to some space-age minds; in any event, recognition of the limitations of the procedure should induce caution in decision making.

*Rotation Time.* Rotation time is the period between successive cuts. One approach in estimating a rotation time that will not diminish long-term productivity of the forest is to estimate the time required for the ecosystem to reacquire nutrients lost as a result of harvesting. Given sufficient time, these losses can be made up by various nutrient inputs into the ecosystem and by the weathering of minerals already in the ecosystem (Chapter 2, Figure 1-14).

Based on our model of ecosystem development, it is apparent that there are several ways in which harvesting practices can alter the length of time necessary to regain nutrients. Using nitrogen as an example, in Figure 8-21 we portray the standing crop in an aggrading forest that has been clear-cut by the stem-only method and allowed to regrow. The curve

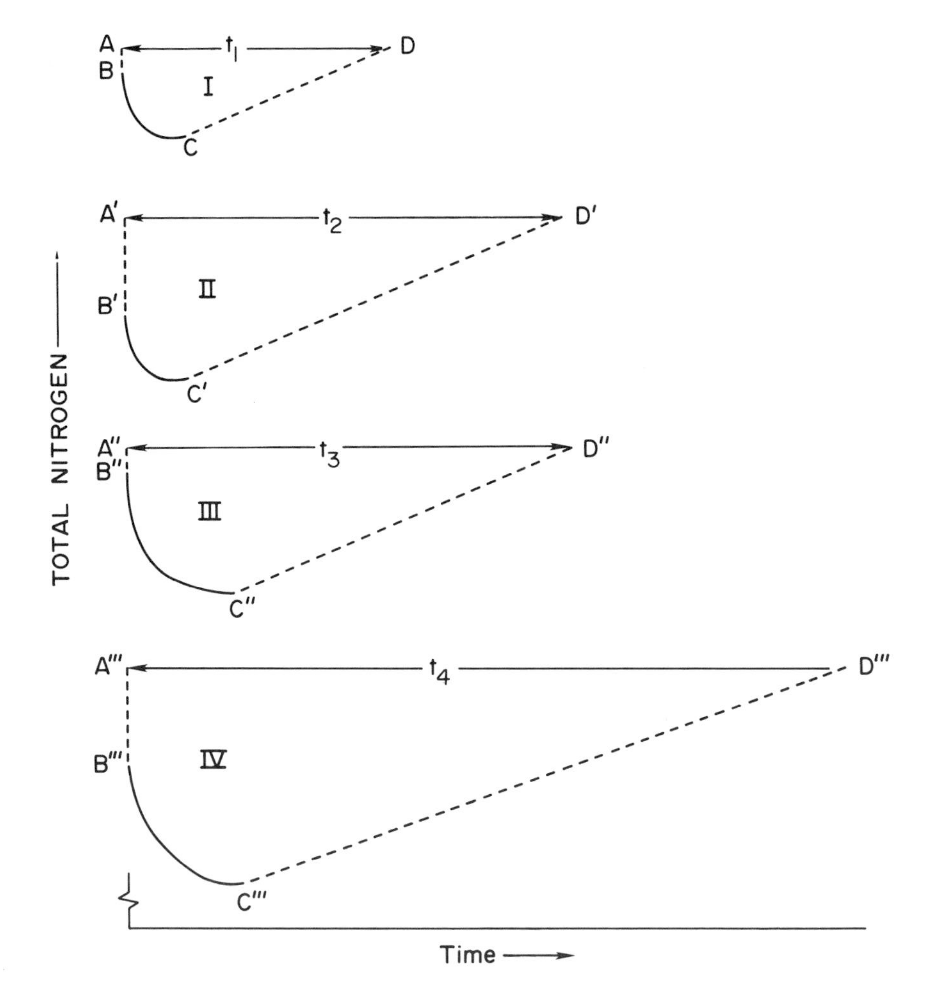

portraying changes in standing crop of nitrogen is broken into three segments—AB, BC, and CD. Segment AB represents nitrogen removed in wood products; BC represents the decline in standing stock during the Reorganization Phase, when losses of dissolved substances and gaseous exports exceed nitrogen inputs in precipitation and nitrogen fixed by microorganisms; CD represents the Aggradation Phase, when inputs exceed outputs and the system accumulates nitrogen (Figure 2-10). The rotation time *t* is the number of years necessary to replenish the standing stock to the precutting level.

Several major points emerge from evaluation of Figure 8-2:

1. The amount of nitrogen lost from the ecosystem is not simply the amount removed in wood products (i.e., segment AB), as suggested by some authors (e.g., Patric and Smith, 1975), but also includes net nitrogen losses during the Reorganization Phase (segment BC). Covington's (1976) estimate of the forest floor decline after stem-only clear-cutting indicates that nitrogen losses from the forest floor may be several times the amount removed in wood harvest. Considering these total losses, it seems clear that estimates of the time necessary to restore nutrients lost during harvesting would be grossly underestimated if based on the replacement of nutrients removed in wood products alone (Figure 8-2I).
2. The amount of nitrogen removed in wood products (segment AB) is strongly affected by the type of harvest. Our estimates at Hubbard

**Figure 8-2.** This figure is intended to illustrate some of the potential effects of various harvesting and seedbed procedures on the time required for northern hardwood ecosystems in the White Mountain region to regain nitrogen lost through harvesting. Time periods are based on hypothetical changes in the standing stock of total nitrogen in the ecosystem. Curve segment AB results from nitrogen removal in wood products; segment BC portrays net nitrogen losses during the Reorganization Phase; Segment CD represents net nitrogen buildup during the Aggradation Phase. The time required by the ecosystem to return to precutting levels of total nitrogen stock, point A, is shown by *t*. (I) Stem-only harvest (AB) with a modest effect on net losses during the Reorganization Phase (BC) and the rate of nitrogen gain during the Aggradation Phase (CD). (II) Complete-tree harvest (A′B′) in which B′C′ = BC, the slope of C′D′ = slope of CD, and $t_2 > t_1$; *t* is primarily influenced by the increased nitrogen removal in wood products. (III) Stem-only harvest (A″B″) in which harvesting procedures increased both nitrogen losses during Reorganization and the length of the Reorganization Phase, i.e., B″C″ > BC, slope of C″D″ = slope of CD, and $t_3 > t_1$; this case suggests that careless harvesting or seedbed practices can substantially lengthen the recovery period. (IV) Complete-tree harvest (A‴B‴) after which net nitrogen losses are increased during the Reorganization Phase and the rate of nitrogen buildup is decreased during the Aggradation Phase, i.e., B‴C‴ > BC, slope of C‴D‴ < slope of CD, and $t_4 >> t_1$. These patterns illustrate that under some circumstances recovery of the ecosystem can be delayed by additive effects on AB, BC, and CD and that the time necessary for ecosystem recovery may be greatly lengthened. Obviously, there are many more possibilities than shown here.

Brook suggest that a complete-tree harvest of a 60-yr-old stand would remove about 4.3 times the nitrogen removed in a stem-only harvest. A whole-tree harvest of a 90-yr-old stand at Bartlett Forest would remove 2.3 times the nitrogen taken out in a stem-only harvest (Hornbeck, 1977). Even if we assumed that all three harvest methods had equivalent losses during the Reorganization Phase (segment BC) and equivalent rates at which nitrogen was accumulated during the Aggradation Phase (segment CD), it is apparent that whole-tree and complete-tree harvests would significantly increase the time necessary to replace nitrogen lost in harvesting (Figure 8-2II).

3. Net nitrogen losses during Reorganization (BC) are probably strongly influenced by management procedures. For example, pulling up root systems in complete-tree harvests or seedbed preparations could result in excessive forest floor disturbance and soil mixing. Such disturbance might lead to increased rates of decomposition and nitrification while reducing the efficiency of natural regeneration in taking up solubilized nutrients. Also, biological nitrogen fixation is known to occur in dead wood, and removal of dead trees, tops, or stumps in whole- or complete-tree harvests might lower the nitrogen input via fixation during reorganization of the ecosystem. As a consequence, net nitrogen losses during Reorganization (BC) would increase and lead to a longer replacement time (Figure 8-2III).
4. Rates of nitrogen accumulation during the Aggradation Phase (segment CD) may be lowered both by the magnitude of harvesting and/or by harvesting procedures. Lowered rates of nitrogen accumulation would lead to an increase in the length of the recovery period (Figure 8-2IV).

This type of analysis suggests that whole-tree or complete-tree harvests do not actually increase production, as proposed by some authors (Lowe, 1973), but rather increase harvests at one time while lengthening the time required for the ecosystem to return to precutting nutrient levels. The return-time may be lengthened simply because more nutrients are removed in wood products, but changes in net nutrient losses during Reorganization and/or decreased rates of accumulation during Aggradation may act, in an additive way, to increase return-times substantially above those calculated on wood-product removal alone.

One important point not shown in Figure 8-2 is that repeated harvests at time periods shorter than those required to regain precutting nutrient levels could cause a general deterioration of the ecosystem. Such deterioration can be thought of as a series of curves such as those shown in Figure 8-2 but with point A at a lower starting level on each successive cut; deterioration also could be represented by effects on the segments BC and/or CD. For example, the rate of nitrogen accumulation (CD) might be lower after each successive cut.

The "puckerbrush" method of complete-tree harvesting may present special problems. Our data indicate that the rapid growth of puckerbrush in White Mountain forests is related to the decline of the forest floor, with some nutrients transferred to the rapidly regrowing vegetation (Chapter 5). It seems likely that repeated harvests at 10- to 20-yr intervals would lead to a rapid decline in forest productivity due to insufficient time to rebuild the forest floor.

It is conceivable that forest fertilization could be used to make up nutrient deficits resulting from more intensive harvesting, but this is not a simple matter of adding a one-time application of fertilizer equivalent to estimated losses. Questions arise as to the amount of nutrients to be replaced, the timing of fertilizer applications (e.g., Waring, 1973), the effect of intensive harvesting on exchange capacity, which in our ecosystems is largely determined by organic surfaces, the amount of fertilizer that moves directly to drainage water (e.g., Ramberg, 1976), and so forth.

Finally, some consideration should be given to the fossil fuel costs attached to new harvesting procedures. In view of the national need to conserve energy, careful evaluations should be made of harvesting techniques that promise more products per unit area but involve energy-expensive machinery and fertilization. An increased use of whole-tree or complete-tree harvesting might represent a shift away from forests that are largely self-maintaining and solar-powered (Bormann, 1976) to forests dependent to a larger extent on fossil energy.

Within this framework, intensive study should be launched to estimate the energy costs, economic benefit, and environmental impact of new harvesting techniques. This should be done immediately, before large amounts of capital are committed to whole-tree, complete-tree, or puckerbrush harvesting and the related processing. Experience from the last decade suggests that once a heavy financial commitment to new technology is made, reasons will be found to continue its use.

## Acceptability of Clear-Cutting

One of the objectives of the Hubbard Brook Ecosystem Study is to link basic research with forest management. To this end we present a few comments on the ecological acceptability of stem-only clear-cutting in northern New England.

Our studies suggest that many similarities exist between redevelopment occurring in clear-cut ecosystems and in openings in the forest created by naturally occurring treefall. This suggests to us that clear-cutting has the potential to work with nature rather than against it and that clear-cutting may be considered as an ecologically acceptable procedure in White Mountain northern hardwood forests. However, it also is apparent that misuse of stem-only clear-cutting can lead to unnecessary short- and long-term degradation of the forest ecosystem. Therefore, it should be coupled with carefully designed safeguards.

In the following paragraphs we suggest some safeguards that have become apparent as a result of the Hubbard Brook Ecosystem Study. By no means are these considered to be all-inclusive, nor are all of the conclusions original with us. Many, but not all, of our suggestions have already been incorporated in stem-only clear-cutting procedures advocated by the U.S. Forest Service.

To be acceptable, clear-cutting (stem-only) systems should meet the following specifications (Likens et al., 1978):

1. Cutting should be limited to sites with strong recuperative capacity. Clear-cutting on steep slopes or on thin soils can lead to long-term changes in the structure, metabolism, and biogeochemistry of the forest ecosystem. This was shown dramatically in the deforestation experiment at Hubbard Brook in which small patches of forest with thin soil on bedrock were converted to bare rock by the accelerated erosion that followed cutting (Reiners et al., 1979).
2. Cutting should be done in the context of a larger watershed unit and in relation to all previous cuts in the unit. This will allow the maintenance of water quality by dilution and by "purifying" activities within the drainage streams.
3. Cuts should be relatively small, for example, several hectares, to insure the availability of seed sources and to minimize losses of dissolved substances and eroded material. Two methods of clear-cutting—block-cutting and progressive strip-cutting—are under study at Hubbard Brook. Block-cutting is a complete clear-cut done all at once. In progressive strip-cutting all trees are harvested over a 4-yr period; the forest is divided into a series of 25-m strips, and every third strip is harvested at 2-yr intervals. Preliminary data (Hornbeck et al., 1975b) indicate that accelerated water and nutrient losses may be significantly reduced by progressive strip-cutting.
4. The cutting and harvesting procedure should do minimum damage to the forest floor. This will safeguard the natural regenerative capacity of the ecosystem and sustain area-wide control over erosion.
5. Roads should consume an absolute minimum amount of area, commensurate with sound ecologic and engineering principles.
6. Mechanical damage to the stream channels should be avoided by leaving a sufficiently wide strip of uncut trees along both banks.
7. Proper ecological weight should be given to species, such as pin cherry, raspberry, and elderberry, which have little importance as a source of wood products. These exploitive species play an important role in the recovery process by conserving nutrients and minimizing erosion and are also an important source of food for wildlife.
8. Planned rotation times should be long enough for the ecosystem to regain, by natural processes, nutrients and organic matter equivalent both to that lost as a result of product removal and to those losses accelerated by clear-cutting. Current studies in the White Mountain

region (Covington, 1976) suggest that, on the average, about 65 years is required for organic matter in the forest floor to rebuild to precutting levels. Hence we suggest that a rotation time in excess of 65 years is compatible with natural regenerative processes. The U.S. Forest Service guidelines for management of timber harvest in the White Mountain National Forest suggest a 110- to 120-yr rotation for cutting (Weingart, 1974).

## LANDSCAPE MANAGEMENT

Plans for the management of landscapes in the White Mountain region include many areas where commercial forestry is very limited or excluded. It is our hope that some of the material in this book and in our earlier volume, *Biogeochemistry of a Forested Ecosystem*, will be useful to landscape planners and managers concerned with those areas.

We suggest in closing that a longer planning frame would allow areas utilized for commercial forestry to serve more effectively a multiplicity of uses and needs, including those of interest to forest industrialists, conservationists, and recreational and regional planners.

If forest cutting could be stabilized so that a certain proportion of all lands dedicated to forest utilization were clear-cut (stem-only) every few years (perhaps on an 80- to 100-yr rotation), taking into account the safeguards mentioned earlier, a number of regional benefits might be obtained. These benefits result from a planned long-term harvesting schedule rather than a boom-or-bust schedule governed solely by market conditions. Such a harvesting regime would create a landscape composed of relatively unchanging proportions of northern hardwood forest in all stages of development. The proportion of the region in each developmental phase would remain constant, but individual sites would undergo development and eventually be cut at the next rotation. The benefits to the region would be as follows:

1. An even flow of wood products might result in more stable wood-using industries.
2. Water flowing from the region would be of high quality, and changes in quality and quantity due to harvesting activities would be minimal.
3. The maintenance of natural gene pools and biological diversity would be enhanced. Species of plants and animals tend to be tied to particular developmental phases of the ecosystem. The cutting regime, along with the maintenance of natural areas not subject to cutting, would guarantee adequate representation of all developmental conditions.
4. Hunting and fishing potential also would be stabilized by maintenance of fairly constant proportions of the landscape in different developmental phases.

5. Although aesthetic and recreational values are largely personal, from several points of view these properties would be enhanced by the increased landscape diversity resulting from availability of all phases of forest growth, optimum diversity of plant and animal species, and streams and lakes of high quality.

Finally, it is necessary to note rapidly changing uses of the forest, other than those related to timber and pulp production. These uses include dispersed and intensive recreation, permanent and vacation housing, increased winter and summer off-road vehicular traffic, and the increased demand for wood as fuel. All of these uses have potential destabilizing effects on the northern hardwood ecosystem.

There is no doubt that the coming decades will see demands for increased yields of goods and services from northern hardwood forests. At the same time, we can be assured that pressures for long-term conservation of our solar-powered forest resources will mount. These diverse goals can be reconciled only through a thorough understanding of the structure, metabolism, and biogeochemistry of natural ecosystems.

# References


Aber, J. D. 1976. A computer model of canopy dynamics and competition for light and nutrients in northern hardwood forests. Ph.D. Thesis, Yale University, New Haven, CT. 136 pp.

Abrahamsen, G., K. Bjor, R. Horntvedt, and B. Tveite. 1976. Effects of acid precipitation on coniferous forest. Pp. 37–63. In: *Impact of Acid Precipitation on Forest and Freshwater Ecosystems in Norway*, F. H. Braekke, ed. SNSF Project, Research Report FR6/76, Oslo, Norway.

Akesson, H. A. 1963. Drainage of forest roads. Canad. Pulp Paper Assoc. Woodlands Sect. Index No. 2260(B-8-b):6–12.

Alexander, M. 1961. *Introduction to Soil Microbiology*. John Wiley, NY. 472 pp.

*Ambio*. 1976. Report from the International Conference on the Effects of Acid Precipitation in Telemark, Norway, June 14–19, 1976. *Ambio*, **5**(5-6):200–264.

Anderson, R. N. 1968. *Germination and Establishment of Weeds for Experimental Purposes*. Weed Sci. Soc. of America. 236 pp.

Anonymous. 1958. *Manual of Forest Fire Control*. Northeast Forest Fire Protection Commission. 268 pp.

Art, H. W. 1976. *Ecological Studies of the Sunken Forest, Fire Island National Seashore*. National Park Service Monograph Series No. 7, NY. 237 pp.

Art, H. W., F. H. Bormann, G. K. Voigt, and G. M. Woodwell. 1974. Barrier Island forest ecosystem: The role of meteorological inputs. *Science*, **184**:60–62.

Baker, F. S. 1949. A revised tolerance table. *J. Forestr.*, **47**:179–181.

Barraclough, S. L. 1949. Forest land ownership in New England with special reference to forest holdings of less than five thousand acres. Ph.D. Thesis, Harvard University, Cambridge, MA. 269 pp.

Barrett, J. W. 1962. *Regional Silviculture of the United States*. Ronald Press, NY. 610 pp.

Belcher, C. F. 1961. The logging railroads of the White Mountains. Part IV. The Zealand Railroad (1884–1892). *Appalachia*, **33**:353–374.

Bjorkbom, J. C. 1967. Seedbed-preparation methods for paper birch. U.S. Forest Service Research Paper NE-79. 15 pp.

Bjorkbom, J. C. 1972. Stand changes in the first 10 years after seedbed preparation for paper birch. USDA Forest Service Research Paper NE-238. 10 pp.

Bloomberg, W. G. 1950. Fire and spruce. *For. Chron.*, **26**:157–161.

Blum, B. M., and S. M. Filip. 1962. A weeding in ten-year-old northern hardwoods—methods and time requirements. USDA Forest Service Northeastern Forest Experiment Station Research Note 135.

Bormann, F. H. 1953. Determining the role of loblolly pine and sweet gum in early old-field succession in the Piedmont of North Carolina. *Ecol. Monogr.*, **23**:339–358

Bormann, F. H. 1976. An inseparable linkage: Conservation of natural ecosystems and the conservation of fossil energy. *BioScience*, **26**:754–760.

Bormann, F. H., and M. F. Buell. 1964. Old-age stands of hemlock northern hardwood forest in central Vermont. *Bull. Torrey Bot. Club*, **91**:451–465.

Bormann, F. H., and G. E. Likens. 1967. Nutrient Cycling. *Science*, **155**:424–429.

Bormann, F. H., G. E. Likens, and J. S. Eaton. 1969. Biotic regulation of particulate and solution losses from a forest ecosystem. *BioScience*, **19**:600–610.

Bormann, F. H., G. E. Likens, and J. M. Melillo. 1977. Nitrogen budget for an aggrading northern hardwood forest ecosystem. *Science*, **196**:981–983.

Bormann, F. H., G. E. Likens, T. G. Siccama, R. S. Pierce, and J. S. Eaton. 1974. The export of nutrients and recovery of stable conditions following deforestation at Hubbard Brook. *Ecol. Monogr.*, **44**:255–277.

Bormann, F. H., and A. P. Nelson. 1963. Vegetation zonation in north-central New England. AIBS Ecol. Soc. Amer. Field Trip Mimeo.

Bormann, F. H., T. G. Siccama, G. E. Likens, and R. H. Whittaker. 1970. The Hubbard Brook Ecosystem Study: Composition and dynamics of the tree stratum. *Ecol. Mongr.*, **40**:377–388.

Botkin, D. B., J. F. Janak, and J. R. Wallis. 1972a. Rationale, limitations, and assumptions of a northeastern forest growth simulator. IBM J. Res. Devel., **16**(2):101–116.

Botkin, D. B., J. F. Janak, and J. R. Wallis. 1972b. Some ecological consequences of a computer model of forest growth. *J. Ecol.*, **60**:948–972; IBM Research Report No. 15799 (1971).

Bourdeau, P. F. 1959. Seasonal variations of photosynthetic efficiency of evergreen conifers. *Ecology*, **40**:63–67.

Brady, N. C. 1974. *The Nature and Properties of Soils*. MacMillan Publishing Co., New York. 639 pp.

Braekke, F. W., ed. 1976. *Impact of Acid Precipitation on Forest and Freshwater Ecosystems in Norway*. SNSF Project Research Report FR6/76. Oslo, Norway. 111 pp.

Braun, E. L. 1950. *Deciduous Forests of Eastern North America*. The Blakiston Co., Philadelphia. 594 pp.

Bray, W. L. 1915. The development of the vegetation of New York State. Coll. Forestry, Syracuse University, Syracuse, NY; Tech. Publ. 3. 186 pp.

Bromley, S. W. 1935. The original forest types of southern New England. *Ecol. Monogr.*, **5**:61–89.

Brown, J. W. 1958a. Forest history of Mount Moosilauke, Part I: Primeval conditions, settlements and the farm–forest economy. *Appalachia* **24**(7):23–32.

Brown, J. W. 1958b. Forest history of Mount Moosilauke, Part II: Big logging days and their aftermath (1890–1940). *Appalachia* **24**(7):221–233.

Bryan, R. W. 1976. New Hampshire crew feeds the whole-tree chipper with feller-buncher, forwarders. *For. Ind.*, **103**(4):48–49.

Campbell, R. W. 1975. The gypsy moth and its natural enemies. Agr. Infor. Bull. No. 381. USDA Forest Service. 27 pp.

Chapman, H. H. 1947. Natural areas. *Ecology*, **28**:193–194.

Chase, A. J., F. Hyland, and H. E. Young. 1971. Puckerbrush pulping studies. Life Sci. and Agric. Exp. Station Tech. Bull. 49, Univ. of Maine, Orono, ME. 64 pp.

Chase, A. J., F. Hyland, and H. E. Young. 1973. The commercial use of puckerbrush pulp. Life Sci. and Agric. Exp. Station Tech. Bull. 65, Univ. of Maine, Orono, ME. 54 pp.

Chase, F. E., C. T. Corke, and J. B. Robinson. 1968. Nitrifying bacteria in soil. Pp. 593–611. In: *The Ecology of Soil Bacteria*, T. R. G. Gray and D. Parkinson, eds. Univ. of Toronto Press, Canada.

Chittenden, A. K. 1905. Forest conditions of northern New Hampshire. USDA Bureau of Forestry Bull. No. 55. 100 pp.

Clark, F. B. 1962. White ash, hackberry, and yellow poplar seed remain viable in the forest litter. *Ind. Acad. Sci. Proc.*, **72**:112–114.

Cleveland, W. S., B. Kleiner, J. E. McRae, and J. L. Warner. 1976. Photochemical air pollution: Transport from the New York City areas into Connecticut and Massachusetts. *Science*, **191**:179–181.

Cline, A. C., and S. H. Spurr. 1942. The virgin upland forest of central New England. Harv. For. Bull. No. 21. 51 pp.

Cogbill, C. V. 1975. Acid precipitation and forest growth in the northeastern United States. M.S. Thesis, Cornell University, Ithaca, NY. 62 pp.

Cogbill, C. V. 1976. The effect of acid precipitation on tree growth in eastern North America. Pp. 1027–1032. In: *Proc. First Internat. Symp. on Acid Precipitation and the Forest Ecosystem*, L. S. Dochinger and T. A. Seliga, eds. USDA. For. Serv. Gen. Tech. Rept. NE-23.

Cole, D. W. and S. P. Gessel. 1965. Movement of elements through a forest soil as influenced by tree removal and fertilizer additions. Pp. 95–104. In: *Forest–Soil Relationships in North America*, C. T. Youngberg, ed. Oregon State Univ. Press, Corvallis, OR.

Colman, E. A. 1953. *Vegetation and Watershed Management.* Ronald Press, NY. 412 pp.

Connell, J. H., and R. O. Slatyer. 1975. Mechanisms of succession in natural communities and their role in community stability and organization. Mimeo., 59 pp.

Coolidge, A. J., and J. B. Mansfield. 1860. *History and Description of New England.* A. J. Coolidge, Boston.

Cooper, W. S. 1913. The climax forest of Isle Royale, Lake Superior, and its development I. *Bot. Gaz.*, **55**:1–44.

Corke, C. T. 1958. Nitrogen transformations in Ontario forest podzols. *North Amer. For. Soils Conf.*, **1**:116.

Covington, W. W. 1976. Forest floor organic matter and nutrient content and leaf fall during secondary succession in northern hardwoods. Ph.D. Thesis, Yale University, New Haven, CT. 98 pp.

Covington, W. W., and J. D. Aber. 1979. Leaf production during secondary succession in northern hardwoods. Submitted to *Ecology*.

Critchfield, W. B. 1960. Leaf dimorphism in *Populus trichocarpa*. *Amer. J. Bot.*, **47**:699–711.

Curry, R. 1971. Soil destruction associated with forest management and prospects for recovery in geologic time. Pp. 157–164. In: *Clearcutting Practices on National Timberlands: Hearings Before the Subcommittee on Public Lands*, 92nd Congress. U.S. Govt. Print. Office, Washington, DC.

Dansereau, P. 1957. *Biogeography: An Ecological Perspective*. Ronald Press, NY. 394 pp.

Davis, R. B. 1966. Spruce–fir forests of the coast of Maine. *Ecol. Monogr.*, **36**:79–94.

Day, G. 1953. The Indian as an ecological factor in the northeastern forest. *Ecology*, **34**:329–346.

Dochinger, L. S., and T. A. Seliga, eds. 1976. *Proceedings of The First International Symposium on Acid Precipitation and the Forest Ecosystem*. USDA Forest Service General Tech. Report NE-23. 1074 pp.

Dominski, A. 1971. Accelerated nitrate production and loss in the northern hardwood forest ecosystem underlain by podzol soils following clear cutting and addition of herbicides. Ph.D. Thesis, Yale University, New Haven, CT. 157 pp.

Douglass, J. E. 1967. Effects of species and arrangement of forests on evapotranspiration. Pp. 451–461. In: *Proc. Internat. Symp. on Forest Hydrology*. Pergamon Press, New York, NY.

Eaton, J. S., G. E. Likens, and F. H. Bormann. 1973. Throughfall and stemflow chemistry in a northern hardwood forest. *J. Ecol.*, **61**:495–508.

Eaton, T. H., Jr., and R. F. Chandler, Jr. 1942. The fauna of the forest humus layers in New York. Cornell Agricultural Experiment Station Memoir 247. 26 pp.

Egler, F. E. 1940. Berkshire plateau vegetation, Massachusetts. *Ecol. Monogr.*, **10**:145–192.

Egler, F. E. 1954. Vegetation science concepts: I. Initial floristic composition a factor in old-field vegetation development. *Vegetatio*, **4**:412–417.

Fernald, M. L. 1950. *Gray's Manual of Botany*. American Book Co., NY. 1632 pp.

Filip, S. M. 1969. Natural regeneration of birch in New England. Pp. 50–54. In: *The Birch Symposium*. USDA Forest Service, Northeastern Forest Experiment Station, Durham, NH.

Filip, S. M., and N. D. Shirley. 1968. Cost of site preparation with a tractor rake in northern hardwood clear cutting. *N. Logger Timber Process.* **17**(5):18–19.

Fisher, S. G. 1970. Annual energy budget of a small forest stream ecosystem: Bear Brook, West Thornton, New Hampshire. Ph.D. Thesis, Dartmouth College, Hanover, NH. 97 pp.

Fisher, S. G., and G. E. Likens. 1972. Stream ecosystem: Organic energy budget. *BioScience*, **22**:33–35.

Fisher, S. G., and G. E. Likens. 1973. Energy flow in Bear Brook, New Hampshire: An integrative approach to stream ecosystem metabolism. *Ecol. Monogr.*, **43**:421–439.

Fobes, C. B. 1948. Historic forest fires in Maine, *Econ. Geogr.*, **24**:269–273.

Forcier, L. K. 1973. Seedling pattern and population dynamics and the reproductive strategies of sugar maple, beech, and yellow birch at Hubbard Brook. Ph.D. Thesis, Yale University, New Haven, CT. 194 pp.

Forcier, L. K. 1975. Reproductive strategies and the co-occurrence of climax tree species. *Science*, **189**:808–810.

Frissell, S. S., Jr. 1973. The importance of fire as a natural ecological factor in Itasca State Park, Minnesota. *Quat. Res.*, **3**(3):397–407.

Fryer, J. H., and F. T. Ledig. 1972. Microevolution of the photosynthetic temperature optimum in relation to the elevational complex gradient. *Canad. J. Bot.*, **50**:1231–1235.

Fryer, J. H., F. T. Ledig, and D. R. Korbobo. 1972. Photosynthetic response of balsam fir seedlings from an altitudinal gradient. Pp. 27–34. In: *Proc. 19th Northeastern Forest Tree Improvement Conference*, Orono, ME.

Galloway, J. N., and G. E. Likens. 1977. Atmospheric enhancement of metal deposition in Adirondack lake sediments. Report to Office of Water Resources Research, Department of the Interior, July 1977. Washington, DC.

Gessel, S. P., D. W. Cole, and E. C. Steinbrenner. 1973. Nitrogen balances in forest ecosystems of the Pacific Northwest. *Soil Biol. Biochem.*, **5**:19–34. Pergamon Press, Great Britain.

Gilbert, A. M., and V. S. Jensen. 1958. A management guide for northern hardwoods in New England. Northeastern Forest Experiment Station Paper No. 112. Upper Darby, PA. 22 pp.

Gilbert, G. K. 1914. *The Transportation of Debris by Running Water.* U.S. Geol. Surv. Prof. Paper 86. 363 pp.

Gimingham, C. H. 1972. *Ecology of Heathlands.* Chapman and Hall, London. 266 pp.

Goodlett, J. C. 1954. Vegetation adjacent to the border of the Wisconsin Drift in Potter County, Pennsylvania. Harvest Forest Bull. No. 25. 91 pp.

Gosz, J. R., R. T. Holmes, G. E. Likens, and F. H. Bormann. 1978. The flow of energy in a forest ecosystem. *Sci. Amer.*, **238**(3):92–102.

Gosz, J. R., G. E. Likens, and F. H. Bormann. 1973. Nutrient release from decomposing leaf and branch litter in the Hubbard Brook forest, New Hampshire. *Ecol. Monogr.*, **43**:173–191.

Gosz, J. R., G. E. Likens, and F. H. Bormann. 1976. Organic matter and nutrient dynamics of the forest floor in the Hubbard Brook forest. *Oecologia (Berl.)*, **22**:305–320.

Gove, W. 1968. The East Branch and Lincoln–A Logger's Railroad. *The Northern Logger and Timber Process.* Pp. 16–17, 48–54.

Graham, S. A. 1941. Climax forests of the upper peninsula of Michigan. *Ecology*, **22**:355–362.

Grime, J. P. 1965. Comparative experiments as a key to the ecology of flowering plants. *Ecology*, **46**:513–515.

Grime, J. P. 1974. Vegetation classification by reference to strategies. *Nature*, **250**:26–31.

Groet, S. S. 1976. Regional and local variations in heavy metal concentrations of bryophytes in the northeastern United States. *Oikos*, **27**:445–456.

Haines, D. A., V. J. Johnson, and W. A. Main. 1975. Wildfire atlas of the Northeastern and North Central States. USDA Forest Service General Technical Report NC-16. 25 pp.

Hale, S. W., et al. 1885. Report of the Forestry Commission of New Hampshire. P. B. Cogswell, Concord, NH.

Hanson, W. A. 1977. Population dynamics, habitat utilization and nutrient cycling of white-tailed deer in the Hubbard Brook Experimental Forest, New Hampshire. (unpublished) 99 pp.

Hardy, R. F. W., R. D. Holsten, E. K. Jackson, and R. C. Burns. 1968. The acetylene-ethylene assay for nitrogen fixation: Laboratory and field evaluation. *Plant Physiol.*, **43**:1185–1207.

Harper, J. L., and J. Ogden. 1970. The reproductive strategy of higher plants: I. The concept of strategy with specific reference to *Senecio vulgaris* L. *J. Ecol.*, **58**:681–689.

Harper, J. L., and J. White. 1974. The demography of plants. *Ann. Rev. Ecol. Systemat.*, **5**:419–463.

Harrington, J. F. 1972. Seed storage and longevity. In: *Seed Biology*, T. T. Kozlowski, ed. Academic Press, New York, NY.

Harries, H. 1965. Soils and vegetation in the alpine and the subalpine belt of the Presidential Range. Ph.D. Thesis, Rutgers University, New Brunswick, NJ. 542 pp.

Hart, G. E., Jr. 1961. Humus depths under cut and uncut northern hardwood forests. USDA Forest Service Research Note 113. 3 pp.

Hart, G. E., Jr., R. E. Leonard, and R. S. Pierce. 1962. Leaf fall, humus depth, and soil frost in a northern hardwood forest. USDA Forest Service, Northeastern Forest Experiment Station, Forest Research Note 131. 3 pp.

Hawley, R. C., and A. F. Hawes. 1912. *Forestry in New England.* John Wiley, New York, NY. 479 pp.

Heinselman, M. L. 1973. Fire in the virgin forests of the Boundary Waters Canoe Area, Minnesota. *Quat. Res.*, **3**(3):329–382.

Henry, J. D., and J. M. A. Swan. 1974. Reconstructing forest history from live and dead plant material—an approach to the study of forest succession in southwest New Hampshire. *Ecology*, **55**:772–783.

Hibbert, A. R. 1967. Forest treatment effects on water yield. Pp. 527–543. In: *International Symposium on Forest Hydrology*, W. E. Sopper and H. W. Lull, eds. Pergamon Press, New York, NY.

Hill, A. R. 1975. Ecosystem stability in relation to stresses caused by human activities. *Canad. Geographer*, **19**(3):206–220.

Hobbie, J. E., and G. E. Likens. 1973. Output of phosphorus, dissolved organic carbon, and fine particulate carbon from Hubbard Brook watersheds. *Limnol. Oceanogr.*, **18**(5):734–742.

Hoeft, R. G., D. R. Keeney, and L. M. Walsh. 1972. Nitrogen and sulfur in precipitation and sulfur dioxide in the atmosphere in Wisconsin. *J. Environ. Qual.*, **1**:203–208.

Holmes, R. T., and F. W. Sturges. 1975. Bird community dynamics and energetics in a northern hardwoods ecosystem. *J. Anim. Ecol.*, **44**:175–200.

Hoover, M. D. 1944. Effect of removal of forest vegetation upon water yields. Trans. Amer. Geophys. Union, Part 6. Pp. 969–975.

Horn, H. S. 1971. *The Adaptive Geometry of Trees.* Princeton University Press, Princeton, NJ. 144 pp.

Hornbeck, J. W. 1973a. The problem of extreme events in paired-watershed studies. USDA Forest Service, Northeastern Forest Experiment Station, U.S. Forest Research Note NE-175. 9 pp.

Hornbeck, J. W. 1973b. Storm flow from hardwood-forested and cleared watersheds in New Hampshire. *Water Resour. Res.*, **9**(2):346–354.

Hornbeck, J. W. 1975. Streamflow response to forest cutting and revegetation. *Water Resour. Bull.*, **11**(6):1257–1260.

Hornbeck, J. W. 1977. Nutrients: A major consideration in intensive forest management. Pp. 241–250. In: *Proceedings of the Symposium on Intensive Culture of Northern Forest Types.* USDA Forest Service General Tech. Rept. NE-29.

Hornbeck, J. W., and C. A. Federer. 1975. Effects of management practices on water quality and quantity: Hubbard Brook Experimental Forest, New Hampshire. Pp. 58–65. In: Municipal Watershed Management Symp. Proc., USDA Forest Service General Tech. Rept. NE-13.

Hornbeck, J. W., R. S. Pierce, and C. A. Federer. 1970. Streamflow changes after forest clearing in New England. *Water Resour. Res.*, **6**(4):1124–1132.

Hornbeck, J. W., G. E. Likens, R. S. Pierce, and F. H. Bormann. 1975a. Strip cutting as a means of protecting site and streamflow quality when clear-cutting northern hardwoods. Pp. 208–229. In: *Proc. 4th North American Forest Soils Conference on Forest Soils and Forest Land Management,* August 1973, B. Bernier and C. H. Winget, eds. Quebec, Canada.

Hornbeck, J. W., R. S. Pierce, G. E. Likens, and C. W. Martin. 1975b. Moderating the impact of contemporary forest cutting on hydrologic and nutrient cycles. Pp. 423–433. In: *Proc. of Internat. Symp. on Hydrologic Sciences.* Publ. 117. Tokyo, Japan.

Horsley, S. B., and H. G. Abbott. 1970. Direct seeding of paper birch in strip clear cutting. *J. Forestr.*, **68**:635–638.

Hough, A. F. 1936. A climax forest community on East Tionesta Creek in northwestern Pennsylvania. *Ecology*, **17**:9–28.

Hough, A. F., and R. D. Forbes. 1943. The ecology and silvics of forests in the high plateaus of Pennsylvania. *Ecol. Monogr.*, **13**:299–320.

Hoyle, M. C. 1965. Growth of yellow birch in a podzol soil. USDA Forest Service Research Paper NE-38. 14 pp.

Hoyle, M. C. 1968. Tree growth and forest soils. Pp. 221–233. In: *Proc. The Third North American Forest Soils Conference,* C. T. Youngberg and C. B. Davey, eds. Oregon State Univ. Press, Corvallis, OR.

Hoyle, M. C. 1973. Nature and properties of some forest soils in the White Mountains of New Hampshire. USDA Forest Service Research Paper NE-260. 18 pp.

Hunt, C. B. 1967. *Physiography of the United States.* Freeman and Co., San Francisco, CA. 480 pp.

Hurst, H. M., and N. H. Burges. 1967. Lignin and humic acids. Pp. 260–286. In: *Soil Biochemistry,* D. A. McLaren and G. H. Peterson, eds. Marcel Dekker, New York, NY.

Hutnik, R. J. 1952. Reproduction on windfalls in a northern hardwood stand. *J. Forestr.*, **50**:693–694.

Jackson, R. M., and F. Raw. 1966. *Life in the Soil.* Institute of Biology's Studies in Biology No. 2. St. Martin's Press, New York, NY. 66 pp.

Jensen, V. S. 1939. Edging white pine lumber in New England. USDA Forest Service, Northeastern Forest Experiment Station Tech. Note No. 26. 2 pp.

Jensen, V. S. 1943. Suggestions for the management of northern hardwood stands in the Northeast. *J. Forestr.*, **41**:180–185.

Johnson, N. M., G. E. Likens, F. H. Bormann, and R. S. Pierce. 1968. Rate of chemical weathering of silicate minerals in New Hampshire. *Geochim. Cosmochim. Acta*, **32**:531–545.

Johnson, N. M., G. E. Likens, F. H. Bormann, D. W. Fisher, and R. S. Pierce. 1969. A working model for the variation in stream water chemistry at the Hubbard Brook Experimental Forest, New Hampshire. *Water Resour. Res.*, **5**(6):1353–1363.

Jones, E. W. 1945. The structure and reproduction of the virgin forest of the North Temperate Zone. *New Phytol.*, **44**:130–148.

Kalm, P. 1770. *Peter Kalm's Travels in North America*, Vol. I. Revised by A. B. Benson, 1937. Dover Publications, New York, NY.

Kelso, E. G. 1969. Birch management of the White Mountain National Forest. Pp. 165–168. In: *The Birch Symposium*. USDA Forest Service, Northeastern Forest Experiment Station, Durham, NH.

Kilgore, B. M. 1973. The ecological role of fire in Sierran conifer forests. *Quat. Res.*, **3**(3):496–513.

Kittredge, J. 1948. *Forest Influences*. McGraw–Hill Book Co., New York, NY. 394 pp.

Kochenderfer, J. N. 1970. Erosion control on logging roads in the Appalachians. USDA Forest Service Research Paper NE-158. 28 pp.

Kochenderfer, J. N. 1977. Area in skidroads, truck roads, and landings in the central Appalachians. *J. Forestr.*, **75**:507–508.

Korstian, C. F. 1924. Growth on cutover and virgin western yellow pine lands in central Idaho. *J. Agri. Res.*, **28**(11):1139–1148.

Kozlowski, T. T., and T. Keller. 1966. Food relations of woody plants. *Bot. Rev.*, **32**:293–382.

Kramer, P. J., and T. T. Kozlowski. 1960. *Physiology of Trees*. McGraw–Hill Book Co., New York, NY. 642 pp.

Küchler, A. W. 1964. Potential natural vegetation of the coterminous United States. Amer. Geogr. Soc. Spec. Publ. No. 36, New York, NY. 116 pp.

Leak, W. B. 1963. Delayed germination of white ash seeds under forest conditions. *J. Forestr.*, **61**:768–772.

Leak, W. B. 1974. Some effects of forest preservation. Northeastern Forest Experiment Station Note NE-186. 41 pp.

Leak, W. B. 1978. Classification of forest habitat in the White Mountains of New Hampshire. 18 pp.

Leak, W. B., and R. E. Graber. 1974. Forest vegetation related to elevation in the White Mountains of New Hampshire. USDA Forest Service. Res. Paper NE-299. 7 pp.

Leak, W. B., and D. S. Solomon. 1975. Influence of residual stand density on the regeneration of northern hardwoods. USDA Forest Service Res. Paper NE-310. 7 pp.

Leak, W. B., D. S. Solomon, and S. M. Filip. 1969. A silvicultural guide for northern hardwoods in the Northeast. USDA Forest Service Res. Paper NE-143. 34 pp.

Leak, W. B., and R. W. Wilson, Jr. 1958. Regeneration after cutting of old-growth northern hardwoods in New Hampshire. USDA Northeastern Forest Experiment Station Paper 103.

Leonard, R. E. 1961. Interception of precipitation by northern hardwoods. USDA Forest Service, Northeastern Forest Experiment Station Paper No. 159. 16 pp.

Leopold, L. B., M. G. Wolman, and J. P. Miller. 1964. *Fluvial Processes in Geomorphology*. W. H. Freeman, San Francisco, CA. 522 pp.

Likens, G. E., F. H. Bormann, and N. M. Johnson. 1969. Nitrification: Importance to nutrient losses from a cut-over forested ecosystem. *Science*, **163**:1205–1206.

Likens, G. E. 1972a. Effects of deforestation on water quality. Pp. 133–140. In: *Proc. Amer. Soc. Civil Engineering Symp. on Inter-disciplinary Aspects of Watershed Management*, Bozeman, MT.

Likens, G. E. 1972b. Mirror Lake: Its past, present and future? *Appalachia*, **39**(2):23–41.

Likens, G. E. 1973. A checklist of organisms for the Hubbard Brook ecosystems. Ecology and Systematics, Cornell Univ., Mimeo. 54 pp.

Likens, G. E. 1976. Acid Precipitation. *Chem. Eng. News*, **54**:29–44.

Likens, G. E., and F. H. Bormann. 1970. Chemical analyses of plant tissues from the Hubbard Brook ecosystem in New Hampshire. Yale University School of Forestry Bull. 79. 25 pp.

Likens, G. E., and F. H. Bormann. 1972a. Biogeochemical cycles. *Sci. Teach.*, **39**(4):15–20.

Likens, G. E., and F. H. Bormann. 1972b. Nutrient cycling in ecosystems. Pp. 25–67. In: *Ecosystem Structure and Function*, J. Wiens, ed. Oregon State Univ. Press, Corvallis, OR.

Likens, G. E., and F. H. Bormann. 1974a. Acid rain: A serious regional environmental problem. *Science*, **184**:1176–1179.

Likens, G. E., and F. H. Bormann. 1974b. Effects of forest clearing on the northern hardwood forest ecosystem and its biogeochemistry. Pp. 330–335. In: *Proc. First Internat. Congress Ecology*, September 1974. Centre Agric. publ. Doc. Wageningen, The Netherlands.

Likens, G. E., and F. H. Bormann. 1974c. Linkages between terrestrial and aquatic ecosystems. *BioScience*, **24**:447–456.

Likens, G. E., F. H. Bormann, N. M. Johnson, and R. S. Pierce. 1967. The calcium, magnesium, potassium, and sodium budgets in a small forested ecosystem. *Ecology*, **48**:772–785.

Likens, G. E., F. H. Bormann, N. M. Johnson, D. W. Fisher, and R. S. Pierce. 1970. Effects of forest cutting and herbicide treatment on nutrient budgets in the Hubbard Brook watershed-ecosystem. *Ecol. Monogr.*, **40**:23–47.

Likens, G. E., F. H. Bormann, R. S. Pierce, J. S. Eaton, and N. M. Johnson. 1977. *Biogeochemistry of a Forested Ecosystem*. Springer-Verlag, New York, NY. 146 pp.

Likens, G. E., F. H. Bormann, R. S. Pierce, and W. A. Reiners. 1978. Recovery of a deforested ecosystem. *Science*, **199**:492–496.

Lorimer, C. G. 1977. The presettlement forest and natural disturbance cycle of northeastern Maine. *Ecology*, **58**:139–148.

Loucks, O. L. 1970. Evolution of diversity, efficiency and community stability. *Amer. Zool.*, **10**:17–25.

Lowe, K. E. 1973. The complete tree: Will it be used to supply the wood fibre needs of the future? *Pulp Paper*, **47**:42–47.

Ludlum, D. M. 1963. *Early American Hurricanes, 1492–1870*. American Meteorological Society, Boston, MA. 198 pp.

Lull, H. W. 1959. Humus depth in the Northeast. *J. Forestr.*, **57**:905–909.

Lull, H. W., and K. G. Reinhart. 1967. Increasing water yield in the Northeast by management of forested watersheds. USDA Forest Service Research Paper NE-66.

Lull, H. W., and K. G. Reinhart. 1972. Forests and floods in the eastern United States. USDA Forest Service Research Paper NE-226. 94 pp.

Lundmark, J. E. 1977. Marken sol del on det skoglia ekosystemet. (The soil as part of the forest ecosystem). *Sv. Skogsvardsforb tidskr,* **75**:109–130.

Lutz, H. J. 1930. The vegetation of Heart's Content, a virgin forest in northwestern Pennsylvania. *Ecology,* **11**:1–29.

Lutz, H. J. 1940. Disturbance of forest soil resulting from uprooting of trees. Yale University School of Forestry Bull. No. 45. 37 pp.

Lutz, H. J., and R. F. Chandler. 1946. *Forest Soils.* John Wiley, New York, NY. 514 pp.

Lyon, C. J., and F. H. Bormann, eds. 1961. *Natural Areas of New Hampshire Suitable for Ecological Research.* Department of Biological Sciences, Publ. No. 2., Dartmouth College, Hanover, NH.

Lyon, C. J., and W. A. Reiners, eds. 1971. *Natural Areas of New Hampshire.* Department of Biological Sciences, Publ. No. 4, Dartmouth College, NH.

MacLean, D. W. 1960. Some aspects of the aspen–birch–spruce–fir type of Ontario. Forest Research Division Tech. Note No. 94. Department of Forestry. 24 pp.

Maissurow, D. K. 1941. The role of fire in the perpetuation of virgin forests of northern Wisconsin. *J. Forestr.,* **39**:201–207.

Margalef, R. 1968. *Perspectives in Ecological Theory.* University of Chicago Press, Chicago, IL. 111 pp.

Margalef, R. 1969. Diversity and stability: A practical proposal and a model of interdependence. Pp. 25–38. In: *Diversity and Stability in Ecological Systems.* Brookhaven Symposia in Biology, No. 22.

Marks, P. L. 1974. The role of pin cherry (*Prunus pensylvanica* L.) in the maintenance of stability in northern hardwood ecosystems. *Ecol. Monogr.,* **44**:73–88.

Marks, P. L. 1975. On the relation between extension growth and successional status of deciduous trees of the northeastern United States. *Bull. Torrey Bot. Club,* **102**:172–177.

Marquis, D. A. 1965. Regeneration of birch and associated hardwoods after patch cutting. USDA Forest Service Research Paper NE-32. 12 pp.

Marquis, D. A. 1967. Clearcutting in northern hardwoods: Results after 30 years. USDA Forest Service Research Paper NE-85. 13 pp.

Marquis, D. A. 1973. The effect of environmental factors in advance regeneration of Allegheny hardwoods. Ph.D. Thesis, Yale University, New Haven, CT. 147 pp.

Marquis, D. A. 1975. Seed storage and germination under northern hardwood forests. *Canad. J. Forestr. Res.,* **5**:478–484.

Marquis, D. A. 1976. Is forestry hurting the forest? An interview. *Green America,* **4**(3). 4 pp.

Marquis, D. A., J. C. Bjorkbom, and G. Yelenosky. 1964. Effect of seedbed condition and light exposure on paper birch regeneration. *J. Forestr.,* **62**:876–881.

Martin, A. C., H. S. Zim, and A. L. Nelson. 1971. *American Wildlife and Plants.* McGraw–Hill, New York. NY. 500 pp.

Mattson, W. J., and N. D. Addy. 1975. Phytophagous insects as regulators of forest primary production. *Science,* **190**:515–522.

McAtee, W. L. 1910. Plants useful to attract birds and protect fruit. USDA Yearbook for 1909. Pp. 185–196.

McIntosh, R. P. 1967. The continuum concept of vegetation. *Bot. Rev.*, **33**:130–187.

Megahan, W. F. 1972. Subsurface flow interception by a logging road in mountains of central Idaho. Pp. 350–356. In: *Proc. of National Symp. Watersheds in Trans.*, Ft. Collins, CO.

Melillo, J. M. 1977. Mineralization of nitrogen in northern forest ecosystems. Ph.D. Thesis, Yale University, New Haven, CT. 136 pp.

Meyer, J. L. 1977. Phosphorus dynamics in a forest stream ecosystem: Less than thirteen in unlucky. *Bull. Ecol. Soc. Amer.* **58**(2):38.

Miller, P. R. 1973. Oxidant-induced community change in a mixed conifer forest Amer. Chem. Soc., Washington, DC; *Adv. Chem. Ser.* **122**:101–117.

Miller, P. R., and M. J. Elderman, eds. 1977. *Photochemical Oxidant Air Pollutant Effects on a Mixed Conifer Forest Ecosystem: A Progress Report.* University of California, Statewide Air Pollution Research Center, Riverside, CA. 338 pp.

Mitchell, H. L., and R. F. Chandler. 1939. The nitrogen nutrition and growth of certain deciduous trees of northeastern United States. The Black Rock Forest Bull. No. 11. 91 pp.

Mitchell, M. J., and D. Parkinson. 1976. Fungal feeding of oribatid mites (*Acari cryptostigmata*) in an aspen woodland soil. *Ecology*, **57**:302–312.

Morey, H. F. 1942. The application of our knowledge of the organic layers of the soil profile to flood control. In: *Proc. of the Hydrology Conference*, F. T. Mavis, ed. School of Engineering Tech. Bull. No. 27. Pennsylvania State College, State College, PA.

Muller, R. N. 1975. The natural history, growth, and ecosystem relations of *Erythronium americanum* Ker. in the northern hardwood forest. Ph.D. Thesis, Yale University, New Haven, CT. 178 pp.

Muller, R. N., and F. H. Bormann. 1976. Role of *Erythronium americanum* Ker. in energy flow and nutrient dynamics of a northern hardwood forest ecosystem. *Science*, **193**:1126–1128.

Mutch, R. W. 1970. Wildland fires and ecosystems: A hypothesis. *Ecology*, **51**:1046–1051.

Nichols, G. E. 1913. The vegetation of Connecticut. *Torreya*, **13**:89–112, 199–215.

Nichols, G. E. 1935. The hemlock–white pine–northern hardwood region of eastern North America. *Ecology*, **16**:403–422.

Nicholson, S. 1965. Altitudinal and exposure variations of the spruce–fir forest on Whiteface Mountain. M.S. Thesis, SUNY, Albany, N.Y. 61 pp.

O'Brien, B. J., and J. D. Stout. 1978. Movement of turnover of soil organic matter indicated by carbon isotope measurements, submitted for publication.

Odum, E. P. 1969. The strategy of ecosystem development. *Science*, **164**:262–270.

Odum, E. P. 1971. *Fundamentals of Ecology*. W. B. Saunders, Philadelphia. PA. 574 pp.

Olmsted, N. W., and J. D. Curtis. 1947. Seeds of the forest floor. *Ecology*, **28**:49–53.

Olsen, R. A. 1956. Absorption of sulphur dioxide from the atmosphere by cotton plants. *Soil Sci.*, **84**:107–111.

Oosting, H. J. 1956. *The Study of Plant Communities*. W. H. Freeman, San Francisco, CA. 440 pp.

Oosting, H. J., and W. D. Billings. 1951. A comparison of virgin spruce–fir forest in the northern and southern Appalachian system. *Ecology*, **32**:84–103.

Parsons, R. F. 1968. The significance of growth-rate comparisons for plant ecology. *Amer. Natur.*, **102**:595–597.

Patric, J. H. 1976a. Is forestry hurting the forest? An interview. *Green America* **4**(3). 4 pp.

Patric, J. H. 1976b. Soil erosion in eastern forests. *J. Forestr.*, **78**:671–677.

Patric, J. H. 1977. Soil erosion and its control in eastern woodlands. *Northern Logger and Timber Process.*, 25(11). 6 pp.

Patric, J. H., and D. W. Smith. 1975. Forest management and nutrient cycling in eastern hardwoods. USDA Forest Service Research Paper NE-324. 12 pp.

Pfister, R. D. 1969. Effect of roads on growth of western white pine plantations in northern Idaho. USDA Forest Service Research Paper INT-65. 8 pp.

Pierce, R. S. 1967. Evidence of overland flow in forest watersheds. Pp. 247–253. In: *Internat. Symp. on Forest Hydrology Proc.*, W. E. Sopper and H. W. Lull, eds. Pergamon Press, Oxford, England.

Pierce, R. S., J. W. Hornbeck, G. E. Likens, and F. H. Bormann. 1970. Effects of elimination of vegetation on stream water quantity and quality. Pp. 311–328. In: *Results on Research on Representative and Experimental Basins*, Proc. of Internat. Assoc. Sci. Hydrology. UNESCO, Wellington, New Zealand.

Pierce, R. S., C. W. Martin, C. C. Reeves, G. E. Likens, and F. H. Bormann. 1972. Nutrient loss from clear-cutting in New Hampshire. Pp. 285–295. In: *Proc. Symp. Watersheds in Trans.*, Ft. Collins, CO.

Pigott, C. D. 1975. Natural regeneration of *Tilia cordata* in relation to forest structure in the forest of Bialowieza, Poland. *Phil. Trans. R. Soc. London Ser. B*, **270**(904):151–179.

Pilgrim, S. L. A., and R. D. Harter. 1977. Spodic horizon characteristics of some forest soils in the White Mountains, New Hampshire. New Hampshire Agricultural Experiment Station Bull. No. 507.

Plummer, F. G. 1912. Forest fires: Their causes, extent and effects with a summary of recorded destruction and loss. USDA Forest Service Bull. 117. 39 pp.

Popay, A. I., and E. H. Roberts. 1970. Factors involved in the dormancy and germination of *Capsella bursa-pastoris* (L.) Medik and *Senecio vulgaris* L. *J. Ecol.*, **58**:103–122.

Ramberg, L. 1976. Effects of forestry operations on aquatic ecosystems. Pp. 143–149. In: *Man and the Boreal Forest*, C. O. Tamm, ed.; *Ecol. Bull., Stockholm* **21**.

Raup, H. M. 1957. Vegetational adjustment to the instability of the site. Pp. 36–48. In: *Proc. and Papers of the 6th Technical Meeting of the Internat. Union for the Protection of Nature.*

Raup, H. M. 1967. American forest biology. *J. Forestr.*, **65**:800–803.

Reiners, W. A., G. E. Likens, F. H. Bormann, and R. S. Pierce. 1979. Vegetational recovery and biogeochemical behavior of a clear-cut watershed ecosystem, ms in preparation.

Reinhart, K. G., A. R. Eschner, and G. R. Trimble, Jr. 1963. Effect on streamflow of four forest practices in the mountains of West Virginia. USDA Forest Service Research Paper NE-1. 79 pp.

*Report of the Chief of the Forest Service: 1970–75.* U.S. Govt. Printing Office, Washington, DC.

Ribe, J. H. 1974. A review of short rotation forestry. Life Sciences and Agricultural Experiment Station, Misc. Report No. 160. University of Maine, Orono, ME. 52 pp.

Rice, E. L. 1974. *Allelopathy*. Academic Press, New York, NY. 353 pp.

Roberts, E. H. 1969. Seed dormancy and oxidative processes. Pp. 161–192. In: *Dormancy and Survival Symposium*, H. W. Woolhouse, ed. Academic Press, New York, NY. 598 pp.

Romell, L. G., and S. O. Heiburg. 1931. Types of humus layer in the forests of northeastern United States. *Ecology*, **12**:567–608.

Roskoski, J. P. 1977. Nitrogen fixation in northern hardwood forests. Ph.D. Thesis, Yale University, New Haven, CT. 112 pp.

Rowe, J. S., and G. W. Scotter. 1973. Fire in the boreal forest. *Quat. Res.*, **3**(3):444–464.

Ryan, D. F. 1978. Nutrient resorption from senescing leaves: A mechanism of biogeochemical cycling in a northern hardwood forest ecosystem. Ph.D. Thesis, Yale University, New Haven, CT.

Safford, L. O. 1974. Effect of fertilization on biomass and nutrient content of fine roots in a beech-birch-maple stand. *Plant Soil*, **40**:349–363.

Safford, L. O., and S. M. Filip. 1974. Biomass and nutrient content of a 4-year-old fertilized and unfertilized northern hardwood stand. *Canad. J. For. Res.*, **4**:549–554.

Salisbury, E. J. 1942. *The Reproductive Capacity of Plants*. G. Bell, London. 244 pp.

Sartz, R. S., and W. D. Huttinger. 1950. Some factors affecting humus development in the Northeast. *J. Forestr.*, **48**:341–344.

Schaller, F. 1968. *Soil Animals*. Ann Arbor Univ. Press, Ann Arbor, MI. 144 pp.

Schlesinger, W. H., W. A. Reiners, and D. S. Knopman. 1974. Heavy metal concentrations and deposition in bulk precipitation in montane ecosystems of New Hampshire, U.S.A. *Environ. Pollut.*, **6**:39–47.

Secrest, H. C., A. J. MacAlong, and R. C. Lorenz. 1941. Causes of the decadence of hemlock at the Menominee Indian Reservation, Wisconsin. *J. Forestr.*, **39**:3–12.

Siccama, T. G. 1971. Presettlement and present forest vegetation in northern Vermont with special reference to Chittenden County. *Amer. Midl. Natl.*, **85**:153–172.

Siccama, T. G. 1974. Vegetation, soil, and climate on the Green Mountains of Vermont. *Ecol. Monogr.*, **44**:325–349.

Siccama, T. G., F. H. Bormann, and G. E. Likens. 1970. The Hubbard Brook Ecosystem Study: Productivity, nutrients, and phytosociology of the herbaceous layer. *Ecol. Monogr.*, **40**:389–402.

Siccama, T. G., and W. H. Smith. 1978. Lead accumulation in a northern hardwood forest. *Environ. Sci. Technol.*, **12**:593–594.

Smith, D. H. 1946. Storm damage in New England forests. M.S. Thesis, Yale University, New Haven, CT. 173 pp.

Smith, F. E. 1975. Ecosystems and evolution. Presidential Address to the Ecological Society of America, Corvallis, OR. 20 pp.

Smith, W. H., F. H. Bormann, and G. E. Likens. 1968. Response of chemoautotrophic nitrifiers to forest cutting. *Soil Sci.*, **106**(6):471–473.

Sopper, W. E., and H. W. Lull. 1965. The representativeness of small forested experimental watersheds in northeastern United States. *Internat. Assoc. Sci. Hydrol.*, **66**(2):441–456.

Sopper, W. E., and H. W. Lull. 1970. Streamflow characteristics of the northeastern United States. Pennsylvania State Univ., University Park, PA. Bull. No. 766. 129 pp.

Spaulding, P., and J. R. Hansbrough. 1944. Decay of logging slash in the Northeast. USDA Forest Service Tech. Bull. No. 876. 22 pp.

Sprugel, D. G. 1976. Dynamic structure of wave-regenerated *Abies balsamea* forests in the northeastern United States. *J. Ecology*, **64**:889–911.

Sprugel, D. G., and F. H. Bormann. 1979. Natural disturbance and the steady-state in high altitude balsam fir forests, ms. in preparation.

Spurr, S. H. 1954. The forests of Itasca in the nineteenth century as related to fire. *Ecology*, **35**:21–25.

Spurr, S. H. 1956a. Natural restocking of forests following the 1938 hurricane in central New England. *Ecology*, **37**:443–451.

Spurr, S. H. 1956b. Forest associations in the Harvard forest. *Ecol. Monogr.*, **26**:245–262.

Spurr, S. H., and B. V. Barnes. 1973. *Forest Ecology*. Ronald Press, New York. 571 pp.

Stearns, F. S. 1949. Ninety years of change in a northern hardwood forest in Wisconsin. *Ecology*, **30**:350–358.

Steinbauer, G. P., and B. Grigsby. 1957. Interaction of temperature, light, and moistening agents in the germination of weed seeds. *Weeds*, **5**:175–182.

Stephens, E. P. 1955. Research in the biological aspects of forest production. *J. Forestr.*, **53**:183–186.

Stephens, E. P. 1956. The uprooting of trees: A forest process. *Soil Sci. Soc. Amer. Proc.*, **20**(1):113–116.

Stevenson, F. J. 1967. Organic acids in soil. Pp. 119–146. In: *Soil Biochemistry*, A. D. McLaren and G. H. Peterson, eds. Marcel Dekker, New York, NY.

Stone, E. 1973. The impact of timber harvest on soils and water. Pp. 427–463. In: *Report of the President's Advisory Panel on Timber and the Environment*. U.S. Govt. Printing Office, Washington, DC.

Striffler, W. D. 1964. Sediment, streamflow and land use relationships in northern lower Michigan. USDA Forest Service Paper LS-16. 12 pp.

Swain, A. M. 1973. A history of fire and vegetation in northeastern Minnesota as recorded in lake sediments. *Quat. Res.*, **3**(3):383–396.

Swank, W. T., and J. E. Douglass. 1974. Streamflow greatly reduced by converting deciduous hardwood stands to pine. *Science*, **185**:857–859.

Swank, W. T., and J. E. Douglass. 1975. Nutrient flux in undisturbed and manipulated forest ecosystems in the Southern Appalachian Mountains. *Publication n° 117 des l'Association Internationale des Sciences Hydrologiques Symposium de Tokyo* (Decembrè 1975).

Swanston, D. N., and F. J. Swanson. 1976. Timber harvesting, mass erosion, and steepland forest geomorphology in the Pacific Northwest. Pp. 199–221. In: *Geomorphology and Engineering*, D. R. Coates, ed. Dowden, Hutchinson and Ross, Stroudsburg, PA.

Tamm, C. O. 1974. Experiments to analyze the behavior of young spruce forests at different nutrient levels. Pp. 266–272. In: *Proc. First Internat. Congress of Ecology: Structure, Functioning and Management of Ecosystems*. Centre for Agr. Publ. and Doc. Wageningen, The Netherlands.

Thompson, D. Q., and R. H. Smith. 1970. The forest primeval in the Northeast—A great myth? Pp. 255–265. In: *Proc. Tall Timbers Fire Ecology Conference No. 10*. Frederickton, New Brunswick, Canada.

Thompson, J. N., and M. F. Willson. 1978. Disturbance and the dispersal of fleshy fruits. *Science*, **200**:1161–1163.

Todd, R. L. 1971. Microbial transformation of nitrogen in natural and manipulated watersheds at Coweeta. EDFB Mimeo. Report No. 71–133.

Trimble, G. R., and H. W. Lull. 1956. The role of forest humus in watershed management in New England. USDA Forest Service Research Paper. 85 pp.

Trimble, G. R., Jr., R. S. Sartz, and R. S. Pierce. 1958. How type of soil frost affects infiltration. *J. Soil Water Conserv.*, **13**(2):81–82.

Trimble, G. R., Jr., and D. W. Seegrist. 1973. Epicormic branching on hardwood trees bordering forest openings. USDA Forest Service Research Paper NE-261. 6 pp.

Tubbs, C. H. 1963. Root development of yellow birch in humus and sandy loam. USDA Forest Service Research Note LS-33. 2 pp.

Ugolini, F. C., H. Dawson, and J. Zachara. 1977. Direct evidence of particle migration in the soil solution of a podzol. *Science*, **198**:603–605.

United States Dept. of Agriculture. 1955. *Wood Handbook*. USDA Handbook No. 72, Washington, DC.

Vitousek, P. M. 1977. The regulation of element concentrations in mountain streams in the northeastern United States. *Ecol. Monogr.*, **47**:65–68.

Vitousek, P. M., and W. A. Reiners. 1975. Ecosystem succession and nutrient retention: A hypothesis. *BioScience*, **25**:376–381.

Walker, J. B., G. B. Chandler, and J. B. Harrison. 1891. *Report of the Forestry Commission of New Hampshire*. J. B. Clarke, Manchester, NH. 53 pp.

Waring, H. D. 1973. Young stands: Early fertilization for maximum production. Presented at FAO/IUFRO Internat. Symp. on Forest Fertilization, Paris.

Watt, A. S. 1925. On the ecology of British beech woods with special reference to their regeneration: II, The development and structure of beech communities on the Sussex downs. *J. Ecol.*, **13**:27–73.

Watt, A. S. 1947. Pattern and process in the plant community. *J. Ecol.*, **35**:1–22.

Wein, R. W., and J. M. Moore. 1977. Fire history and rotations in the New Brunswick Acadian Forest. *Canad. J. For. Res.*, **7**:285–294.

Weingart, P. D. 1974. Forest plan: White Mountain National Forest. Eastern Region, USDA Forest Service. 77 pp.

Wendel, G. W. 1972. Longevity of black cherry seed in the forest floor. USDA Forest Service Research Note NE-149. 4 pp.

Whitney, G. G. 1976. The bifurcation ratio as an indicator of adaptive strategy in woody plant species. *Bull. Torrey Bot. Club*, **103**:67–72.

Whitney, G. G. 1978. A demographic analysis of *Rubus idaeus* L. and *Rubus pubescens* Raf.: The reproductive traits and population dynamics of two temporally isolated members of the genus *Rubus*. Ph.D. Thesis, Yale University, New Haven, CT. 139 pp.

Whittaker, R. H. 1967. Gradient analysis of vegetation. *Biol. Rev.*, **42**:207–264.

Whittaker, R. H. 1975. *Communities and Ecosystems*. MacMillan, New York, NY. 385 pp.

Whittaker, R. H., F. H. Bormann, G. E. Likens, and T. G. Siccama. 1974. The Hubbard Brook Ecosystem Study: Forest biomass and production. *Ecol. Monogr.*, **44**:233–254.

Whittaker, R. H., G. E. Likens, F. H. Bormann, J. S. Eaton, and T. G. Siccama. 1978. The Hubbard Brook Ecosystem Study: Forest nutrient cycling and element behavior. *Ecology*, in press.

Whittaker, R. H., and G. M. Woodwell. 1967. Surface area relations of woody plants and forest communities. *Amer. J. Bot.*, **54**(8):931–939.

Whittaker, R. H., and G. M. Woodwell. 1972. Evolution of natural communities. Pp. 137–156. In: *Ecosystem Structure and Function*, J. A. Wiens, ed. Oregon State Univ. Press, Corvallis, OR.

Williams, W. T., M. Brady, and S. C. Willison. 1977. Air pollution damage to the forests of the Sierra Nevada Mountains of California. *APCA Jour.*, **27**(3):230–234.

Wilson, R. W., and V. S. Jensen. 1954. Regeneration after clear-cutting second-growth northern hardwoods. USDA Forest Service Research Note 27. 3 pp.

Winer, H. I. 1955. History of the Great Mountain Forest. Ph.D. Thesis, Yale University, New Haven, CT. 278 pp.

Wood, T., and F. H. Bormann. 1974. The effects of an artificial acid mist upon the growth of *Betula alleghaniensis* Britt. *Environ. Pollut.*, **7**:259–268.

Woodwell, G. M. 1974. Success, succession and Adam Smith. *BioScience*, **24**:81–87.

Woodwell, G. M., and A. H. Sparrow. 1965. Effects of ionizing radiation on ecological systems. Pp. 20–38. In: *Ecological Effects of Nuclear War*, G. M. Woodwell, ed. Brookhaven National Laboratory 917 (C-43).

Wright, H. E., Jr., and M. L. Heinselman. 1973. The ecological role of fire in natural conifer forest of western and northern North America. *Quat. Res.*, **3**(3):319–328.

Young, H. E. 1974. Biomass, nutrient elements, harvesting and chipping in the complete tree concept. *Consultant*, **19**:91–104.

Zasada, Z. A. 1974. Effects of mechanical harvesting upon the forest ecosystem and silvicultural practices in the northern lake states. Pp. 337–348. In: *Forest Harvesting Mechanism and Automation*, Proc. IUFRO, Division 3, Canad. Forestry Service, Dept. of Environment, Ontario, Canada. Publ. 5.

# Index

*An Interdisciplinary Journal from Springer-Verlag*

## Environmental Management